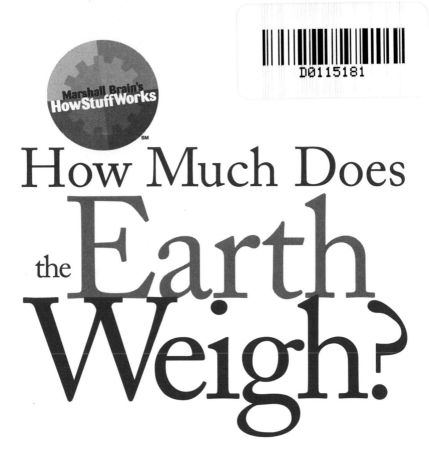

Marshall Brain's
HowStuffWorks ℠

How Much Does the Earth Weigh?

Hungry Minds™

Best-Selling Books • Digital Downloads • e-Books • Answer Networks • e-Newsletters • Branded Web Sites • e-Learning

New York, NY • Cleveland, OH • Indianapolis, IN

Hungry Minds™

Published by
Hungry Minds, Inc.
909 Third Avenue
New York, NY 10022
www.hungryminds.com

For general information on Hungry Minds' products and services please contact our Customer Care Department within the U.S. at 800-762-2974, outside the U.S. at 317-572-3993 or fax 317-572-4002.

For sales inquiries and reseller information, including discounts, premium and bulk quantity sales, and foreign-language translations, please contact our Customer Care Department at 800-434-3422, fax 317-572-4002, or write to Hungry Minds, Inc., Attn: Customer Care Department, 10475 Crosspoint Boulevard, Indianapolis, IN 46256.

Library of Congress Cataloging-in-Publication Data available from the publisher.
ISBN: 0-7645-6519-2

How Stuff Works, Inc.

Editor in Chief: Marshall Brain
Editorial Director: Katherine Fordham Neer
Art Director: Rick Barnes
Contributing Writers: Marshall Brain, Kevin Bonsor, Craig Freudenrich, Ph.D., Tom Harris, Karim Nice, and Jeff Tyson
Production Assistant: Sally Guaspari

Hungry Minds, Inc.

Vice President and Publisher: Kathy Nebenhaus
Editorial Director: Cindy Kitchel
Creative Director: Michele Laseau
Special Editorial Assistance: S. Kristi Hart, Brian Hahn

Cover and Layout Design: Michele Laseau, HMI; Rick Barnes, HSW; Cynthia Anderson, Studio Alchemy; and Alex Mabilon
For more cool questions and answers visit www.howstuffworks.com
Manufactured in the United States of America.
10-9-8-7-6-5-4-3-2-1

How Much Does the Earth Weigh?

⚙ Table of Contents

	Introduction	xiii
Chapter 1	Now That's Entertainment!	1
Chapter 2	Special Effects	11
Chapter 3	Light and Sound	21
Chapter 4	But I Thought GIF was Peanut Butter	33
Chapter 5	On the Go	45
Chapter 6	Power Up!	59
Chapter 7	Stuff Around Your House	71
Chapter 8	Foodstuffs	81
Chapter 9	To Your Health	93
Chapter 10	Who's On First?	105
Chapter 11	Cool Problems/Cool Solutions	119
	Index	133

Chapter **1** **Now That's Entertainment!** **1**

How do the light sabers in the *Star Wars* movies work?

How do fog machines work?

How do light guns work on a video game?

How do they create the special effects in movies like *The Matrix* and in commercials where the camera rotates around a frozen actor?

How is sound recorded on motion picture films?

What does the clapperboard in movie production do?

On television, how does closed captioning work?

Why do all FM radio stations end in an odd number, such as 105.7, 93.3, and 96.9?

Chapter **2** **Special Effects** **11**

Why do Wint-O-Green LifeSavers throw out sparks when you bite them in the dark?

How do trick birthday candles work — the kind that relight after you blow them out?

How does dry ice work?

How do mood rings work, and do they really tell my mood?

How do scratch-and-sniff stickers work? What makes them last for years and years?

How does Alka-Seltzer work and why does it fizz?

How does Pop Rocks candy work?

How do film companies color old black-and-white movies?

Chapter **3** **Light and Sound** **21**

Why is the sky blue?

Could I see a flashlight beam from the earth on the moon?

How do a zebra's stripes act as a camouflage since zebras do not live in a black-and-white environment?

What causes a sonic boom?

What is a light year?

How does a gun silencer work?

What is a decibel?

How do motion sensors work?

How does feedback work in a PA system to cause that howling sound?

How does an Indiglo watch work?

Chapter 4 **But I Thought GIF was Peanut Butter...**   **33**

Does adding more RAM to my computer make it faster?

What do the different Web page extensions mean (html, htm, asp, and so on)?

What is Linux?

Is it better to turn my computer off when I am not using it, or to leave it on all the time?

What is a T1 line?

Why are there so many different image formats on the Web? For example, what is the difference between a GIF and a JPG image?

What is the fastest computer in the world?

What is the Year 2038 Problem?

Chapter 5 **On the Go**   **45**

What is the difference between gasoline, kerosene, diesel fuel, and other forms of oil?

What is the difference between a turbocharger and a supercharger on a car's engine?

How does a Jake Brake work on a big rig?

What does the "weight" mean on a can of motor oil?

How does a car's rearview mirror work when it's set on the upward, glare-resistant setting?

How does a traffic light detect that a car has pulled up and is waiting for the light to change?

How does a gasoline pump at a filling station know when the tank is full?

Why do some engines use a dry sump oil system?

If diesel engines are more efficient and use cheaper fuel than traditional gasoline engines, why don't all cars use diesel engines?

Is it true that a diesel engine can operate under water while a gasoline engine cannot?

When I go to the gas station, I always have a choice of three different octanes: 87, 89, and 93. What is the difference, and what does "unleaded" mean?

Chapter 6 Power Up!   59

Is it possible to generate electricity from heat?

Is flour flammable? I heard if you were to burn flour, it would explode. If so, why does this happen?

How does an oxygen canister work?

What home appliances use the most power?

Do commercial jets have locks on the doors and ignition keys?

My lawn mower doesn't have a battery like a car does, so where does the electricity to spark the spark plug come from?

How does a laser speed gun work to measure a car's speed compared to normal police radar?

How many solar cells would it take to provide the electricity for my house?

How much coal would it take to light a 100-watt light bulb 24 hours a day for 1 year?

How are torpedoes propelled through the water?

Chapter 7   **Stuff Around Your House**   **71**

How do pop-up turkey timers work?

What is this bumpy stuff on my ceiling that looks like popcorn or cottage cheese?

How do electric stud finders know the location of the studs?

How does pressure-treated lumber work? What does "pressure-treated" mean?

Why do the two flat prongs on the plugs for electrical appliances have holes in them?

What is the difference between analog and digital cell phones?

How does caller ID work?

How do the lamps that you can touch to turn them on work?

How do radio signals from the National Atomic Clock in Colorado reach mine?

Chapter 8   **Foodstuffs**   **81**

How does popcorn work?

How much sugar is in soft drinks?

Why is root beer called root beer?

How is cotton candy made?

What is carrageenan?

Why shouldn't dogs eat chocolate?

How does that a plastic ball-shaped widget inside beer release gas to aerate the beer?

How is the caffeine removed from coffee beans?

What is mayonnaise and how is it made?

Chapter 9   **To Your Health**   **93**

Why does the hair on my arms stay short, while the hair on my head can grow very long?

Why does hydrogen peroxide foam when I put it on a cut?

What does it mean when someone has 20/20 vision?

I was working in my garden and got a horrible case of poison ivy. What exactly causes this reaction?

How many senses does a person have? I always hear about five: touch, taste, smell, vision, and hearing. Do we have more?

What makes knuckles pop?

What constitutes a person's IQ? Does it improve with maturity, education, and experience?

How many calories do does a person need daily?

What causes flatulence?

How does your stomach keep from digesting itself?

Chapter **10** Who's on First? **105**

Baseball fields often have checkerboards and other patterns mowed into them. How can I create the same effect with my lawn?

In baseball, how does a pitcher throw a curveball?

In scuba diving, what causes "the bends?"

How is the ice in ice skating rinks made? And how are the logos and lines put on the ice?

How is the first down line superimposed onto the field on televised football games?

Why do golf balls have dimples?

Why does the grass on the greens at a golf course look so perfect? Could my lawn look like this?

How does the ball return work on a coin-operated pool table?

Why shouldn't I go swimming right after I eat?

What makes NASCAR engines different from the engines in street cars?

What would happen if I drilled a tunnel to the center of the earth and jumped into it?

I often hear the expression, "If I had all the money in the world…" So, how much is all the money in the world?

Can a telescope detect the equipment left behind by astronauts on the moon to prove/disprove missions?

How much ice would I have to store up in the winter if I wanted to air condition my house all summer?

How many sheets of paper can be produced from a single tree?

If you took all the matter in the universe and pushed it into one corner, how much space would it take up?

Is there a way to actually see a satellite in orbit?

If daytime running lights were mandatory in the United States and all vehicles had them, how much extra gasoline would be used each year?

If you could build a train that could travel as fast as a bullet, what would happen if you fired a gun from the back or from the front of the train?

How much does the earth welgh?

Index   **133**

Hello!

Readers send hundreds of questions to us at How Stuff Works every day. We answer many of them by email and in the forums. But we pick one especially good question to publish as the official "Question of the Day" each morning.

The question of the day is a long tradition at How Stuff Works, and it is one of the most popular features of the site. We have a huge archive containing the hundreds of questions we have answered since the How Stuff Works Web site started, and millions of people visit the archive to find answers to their most burning questions.

In this book, I've collected over 100 of the most popular questions in the archive. These are the best of the best—the questions that readers just HAVE to know the answer to. They are truly fascinating to read through because these are the most intriguing questions on the site.

As you look through the book, you will notice two icons.

 One is the Top 20 icon. It identifies the 20 most popular questions that How Stuff Works has ever covered.

 Then there is the MB icon. These are my personal favorites. I don't know why, but these questions completely fascinated me or blew me away when I discovered the answer.

I'm an extremely curious person. I'm constantly trying to unearth the mysteries behind the things around me. If you, too, find yourself wondering about, well, just about everything, you will love this collection of questions. And if you have other questions, come to How Stuff Works and ask away!

Now That's Entertainment!

⚙ How do the light sabers in the *Star Wars* movies work? • How do fog machines work? • How do light guns work on a video game? • How do they create the special effects in movies like *The Matrix* and in commercials where the camera rotates around a frozen actor? • How is sound recorded on motion picture films? • What does the clapperboard in movie production do? • On television, how does closed captioning work? • Why do all FM radio stations end in an odd number, such as 105.7, 93.3, and 96.9?

How do the light sabers in the *Star Wars* movies work? MB

Like the *Millennium Falcon*, Yoda, and even *Chitty Chitty Bang Bang* for that matter, a light saber is a special effect that looks so real you actually believe it exists.

The technique used to create the light saber effect in the original *Star Wars* film is straightforward but tedious. On the set, the actors use light sabers made of handles attached to aluminum rods of the correct length. The handles are plastic models and the aluminum rods are painted red or green or blue. The actors use these props as though they were light sabers.

After the film is shot, it is taken to the special effects department. In this initial film, the actors look like they are fighting with painted broomsticks instead of light sabers. A special-effects artist now has the job of making those broomsticks look real. The artist looks at the film frame by frame and projects each frame that contains a light saber onto a clear piece of plastic (an animation cel). The special-effects artist draws the outline of each light saber blade in the frame onto the cel. Then, for each frame, the artist paints in the correct color for the blade using a bright cartoon color. Eventually, the artist has a stack of these cels, one for each frame of the movie containing a light saber. The cels are clear everywhere except where the light saber blade is seen in each frame.

Now a new piece of movie film is shot. On this film each animation cel is placed over a black background and shot with a light diffuser over the lens; this diffuser gives the light sabers the glow they have around the edges. If you were to play this film in a projector, all you would see is the light saber blades moving on a black background. Before it is developed, however, the actual footage from the movie is double-exposed onto this same film. The effect is amazing: The light sabers look bright and real!

As movies move more into the digital realm, the job of animating the light sabers gets slightly easier — but not much. In a digital world each frame of the movie is scanned into a computer at extremely high resolution so that each frame can be manipulated on a computer screen. To make the light sabers look real, the special-effects artist looks at each frame on the computer screen, outlines the broomsticks, colors the areas, and diffuses them (frame by frame by frame…). Instead of being done on a plastic cel, it is all done on separate "cels" in the computer's memory and then merged digitally. The animator must still look at each frame, however, and tediously outline the light saber blades one by one.

○ **Web Links**

How Blue Screen Special Effects Work

How do fog machines work? ◣

There are three common ways to produce the fog that you often see in stage productions and at dance clubs:

- Use a fog machine that vaporizes "fog juice"
- Use dry ice
- Invite lots of cigarette smokers

Fog machines and fog juice are the most common methods. The basic mechanism is simple: The juice is heated to create smoke. When you overheat oil on the stove and create a lot of smoke, you are doing approximately the same thing. Cooking oil has a tendency to get gummy and smell bad, however. Fog machines therefore use glycerin or glycol mixed with water, such as propylene glycol and triethylene glycol mixed with 20% water. It is not known whether this fog has any side effects on people's lungs. The fog seems to be problematic for asthmatics, but nothing has been proven conclusively for the general population.

If you would rather make completely safe fog, you can use dry ice. (To find a source for dry ice, look up "ice" in the business directory — and never handle the stuff with bare hands!). When you place dry ice in hot water it creates a dense fog that clings to the floor. This fog contains carbon dioxide (CO_2)and water vapor; you want to be sure the room is ventilated so that CO_2 build-up is not a problem.

○

How do light guns work on a video game?

Most home video games and many arcade games can use some sort of gun as an input device. You point the gun at the screen and pull the trigger, and if you hit the target on the screen, the target explodes.

To create this effect, the gun contains a photodiode (or a phototransistor) in the barrel. The photodiode is able to sense light coming from the screen. The gun also contains a trigger switch. The output of the photodiode and the switch are fed to the computer controlling the game.

The computer receives signals from the screen driver electronics. The screen driver electronics send pulses to the computer at the start of the horizontal and vertical retrace signals so that the computer knows where the electron beam is on the screen during each frame. The computer normally uses one of two different techniques to determine whether or not the gun is pointed at the target when the user pulls the trigger:

- When the user pulls the trigger, the computer blanks the screen for one frame and then paints just the target object on the screen (as a white object). If the photodiode senses darkness after one vertical retrace signal and light after the next, the computer assumes that the gun is pointed at the target of the screen and scores a hit.

- The computer can blank the screen and then paint the entire screen white. It will take time for the electron beam to trace the entire screen while painting it white. By comparing the signal coming from the photodiode with the horizontal and vertical retrace signals, the computer can detect where the electron beam is on the screen when the photodiode first senses its light. The computer counts the number of microseconds that pass between the time the horizontal and vertical retrace signals start and the photodiode first senses light. The number of microseconds tells the computer exactly where the gun points on the screen. If the calculated position and the position of the target match, the computer scores a hit.

When you pull the trigger, you can normally see the display flash. From the type of flash you see, you can figure out which system is used.

○ Web Links

How Television Works

How do they create the special effects in movies like *The Matrix* and in commercials where the camera rotates around a frozen actor?

This effect is fascinating to watch! In one commercial a horse stops in mid-air and the camera pans around it. In *The Matrix* the technique is used four times only but is so startling that it leaves an impression over the entire movie.

In the commercials, a simpler technique is used than was used in the film. A collection of perhaps 30 still cameras is set up around the object. At the moment where the action should freeze, all 30 cameras fire at once. The images are played one after another to show the rotation.

In *The Matrix*, the filmmakers used an extremely sophisticated technique to accomplish much more advanced effects. In this technique, not only does the rotation occur, but the actor is also moving in slow motion during the rotation. At least five different special-effect techniques combine to create the final image:

- A large number of still cameras capture the scene, but they fire sequentially around the actor rather than all at once.
- The cameras shoot the actor on a green-screen background.
- The actor is wearing a wire suspended from the ceiling so that he can fall only partway or appear to float in mid-air.
- After the scene is shot, software similar to morphing software interpolates between the images to allow the slow-motion feel. The filmmaker can therefore slow down or speed up the action at will.

Now That's Entertainment!

5

- Finally, computer-generated backgrounds are superimposed onto the film.

A technician deals with the film one image at a time using a computer and digitized versions of the images. Once the still images are perfect, the morphing software interpolates between them. Then the background images are laid into the green area. A technician has to build a complete 3-D computer model of the computer-generated scene and then key the rotation through this scene to the position of the camera in each frame of the film.

⚙ Web Links

How Blue Screen Special Effects Work

How is sound recorded on motion picture films?

Depending on where and when you went to high school, you may remember the teacher wheeling in the 16mm film projector to play films in class. If you ever picked up a strip of the film and looked at it along the sprocket holes, you would have seen a dark strip with a wavy white line down the middle. This strip holds an optical soundtrack for the film.

The white line running down the dark strip varies in width. A lamp shines through one side of the strip and hits a photo-detector on the other side. The photodetector's output runs through an amplifier and drives a speaker. The vibrations of the sound are translated into changes in the white strip's width. The photo-detector sees any variations in light intensity and reproduces the sound. The optical sound system is easy to add to the film and is reliable over the life of the film.

This same optical system was used in theaters playing 35mm films when talking movies were first released. In the 1950s, optical soundtracks were replaced by magnetic recording; magnetic strips, just like those on a cassette tape, were applied to the film and sound was recorded on those strips. Magnetic recording allowed for stereo sound and surround sound, and also improved the sound quality of the films. Unfortunately, there were problems with magnetic strip life, and using the magnetic strips made the films a lot more expensive.

In the 1970s, Dolby Laboratories made possible stereo sound using two optical tracks. This system allows for stereo playback and surround sound, and offers Dolby noise reduction as well. Dolby Laboratories has been improving the system for many years.

⚙ Web Links

How Tape Recorders Work

What does the clapperboard in movie production do?

If you have a video camcorder, you are used to video and sound getting recorded on the same tape. The sound and video are always synchronized because they are recorded together in one place.

When filming a movie, however, the pictures and sound are recorded separately. The picture is recorded on film in the camera, and the sound gets recorded on a separate analog magnetic tape recorder (or, more recently, on digital tape such as a DAT tape). Because the picture and sound are recorded on two different devices, you need a way to synchronize them.

A clapperboard is the traditional way to handle the synchronization. The bottom of the clapperboard is normally a slate of some sort on which you can write the scene and take number. This information helps identify the shot during editing. Once the tape recorder and camera are rolling, the clapperboard operator places the clapperboard in front of the camera so that the camera can see it, reads the scene and take information so that the tape recorder can hear it, and then claps the clapper. During editing, it is easy to synchronize the visual of the clapper clapping with the "clap" sound it makes on the tape.

The digital slate is the more modern form of the clapperboard. The tape recorder contains a timecode generator. The timecode is recorded continuously on a special track on the tape, and the timecode is also displayed continuously on a large LED display on the digital slate. By showing the digital slate to the camera before the action starts, the editor knows exactly what the tape's timecode is and can synchronize it with the film.

⚙ Web Links

How Television Works

On television, how does closed captioning work?

Closed captioning can be extremely helpful in at least three different situations:

- It has been a great boon to hearing-impaired television viewers.
- It can be helpful in noisy environments. For example, a TV in a noisy airport terminal can display closed captioning and still be usable.
- Some people use captions to learn English or learn to read.

Closed captioning is embedded in the television signal and becomes visible when you use a special decoder. The decoder lets viewers see captions, usually at the bottom of the screen, which will tell them what is being said or heard on TV shows. Since 1993, television sets with screens of 13 inches or more that are sold in the U.S. must have built-in decoders, under the Television Decoder Circuitry Act. Set-top decoders are available, too, for older TV sets.

The captions are hidden in the line 21 data area found in the vertical blanking interval of the television signal. The blanking interval is the area of the television signal that tells the electron gun to shoot back up to the upper left corner of the screen to begin painting the next frame. Line 21 is the line in the vertical blanking interval that has been assigned to captioning (as well as time and V-chip information). Each frame of video can transmit two characters of captioning information (or special commands that control color, pop-ups, and so on.)

Many shows and commercials now also carry captions. In addition, captions are often added for the reruns of older programs made before captioning became widespread.

Some shows are captioned in real time. That is, during a live broadcast of a special event or of a news program, captions appear just a few seconds behind the action to show what is being said. A stenographer listens to the broadcast and types the words into a special computer program that adds the captions to the television signal. The typists have to be skilled at dictation and spelling and they have to be very fast and accurate at typing.

⚙ **Web Links**

How Does A V-Chip Work?
How Television Works

Why do all FM radio stations end in an odd number, such as 105.7, 93.3, and 96.9?

The Federal Communications Commission (FCC) allocates different frequencies to different activities in the U.S. For example, cellular phones have their own assigned frequencies, baby monitors have their own frequencies, CB radios have their own, and so on.

FM radio stations all transmit in a band between 88 megahertz (millions of cycles per second) and 108 megahertz. This band of frequencies is based mostly on history and whim. Inside that band, each station occupies a 200-kilohertz slice, and all of the slices start on odd number boundaries. So there can be a station at 88.1 megahertz, 88.3 megahertz, 88.5 megahertz, and so on. The 200-kilohertz spacing, and the fact that all the stations are on odd boundaries, is again completely arbitrary. For example, in Europe, the FM stations are spaced 100 kilohertz apart instead of 200 kilohertz apart, and they can be even or odd.

Now That's Entertainment!

⚙

⚙ **Web Links**   **9**

How the Radio Spectrum Works
How Radio Works

Special Effects

✿ Why do Wint-O-Green LifeSavers throw out sparks when you bite them in the dark? • How do trick birthday candles work — the kind that relight after you blow them out? • How does dry ice work? • How do mood rings work, and do they really tell my mood? • How do scratch-and-sniff stickers work? What makes them last for years and years? • How does Alka-Seltzer work and why does it fizz? • How does Pop Rocks candy work? • How do film companies color old black-and-white movies?

Why do Wint-O-Green LifeSavers throw out sparks when you bite them in the dark?

Actually, all hard sugar-based candies emit some amount of light when you bite them, but most of the time, that light is very faint. This effect is called *triboluminescence* and is similar to the electrical charge build-up that produces lightning — only much less grand. Triboluminescence is the emission of light resulting from something being smashed or torn. When you rip a piece of tape off the roll, it will produce a slight glow for the same reason.

Triboluminescence occurs when molecules, in this case crystalline sugars, are crushed, forcing some electrons out of their atomic fields. These free electrons bump into nitrogen molecules in the air. When they collide, the electrons impart energy to the nitrogen molecules, causing them to vibrate. In this excited state, and in order to get rid of the excess energy, these nitrogen molecules emit light — mostly ultraviolet (nonvisible) light, but they also emit a small amount of visible light. This is why all hard, sugary candies will produce a faint glow when cracked.

When you bite into a Wint-O-Green LifeSaver, however, a much greater amount of visible light can be seen. This brighter light is produced by the wintergreen flavoring. Methyl salicylate, or oil of wintergreen, is fluorescent, meaning it absorbs light of a shorter wavelength and then emits it as light of a longer wavelength. Ultraviolet light has a shorter wavelength than visible light. So when a Wint-O-Green LifeSaver is crushed between your teeth, the methyl salicylate molecules absorb the ultraviolet, shorter wavelength light produced by the excited nitrogen. This light is then re-emitted as light of the visible spectrum, specifically as blue light — thus the blue sparks that jump out of your mouth when you crunch on a Wint-O-Green LifeSaver.

How do trick birthday candles work — the kind that relight after you blow them out?

If you have ever seen trick birthday candles, you know that they work amazingly well! When a person blows one out, it simply relights itself in a few seconds.

To understand trick birthday candles, it is helpful to first understand normal candles. The key to a trick candle happens the moment you blow it out. With a normal candle, a burning ember in the wick causes a ribbon of paraffin smoke to rise from the wick. That ember is hot enough to vaporize paraffin but is not hot enough to ignite the paraffin vapor.

The key to a relighting candle, therefore, is to add something to the wick that the ember is hot enough to ignite. The ember can then ignite this substance and the substance can in turn ignite the paraffin vapor. The most common substance used is magnesium. Magnesium is a metal, but it happens to burn (combine with oxygen to produce light and heat) rapidly at an ignition temperature as low as 800°F (430°C). Aluminum and iron both burn as well, but magnesium lights at a lower temperature.

Inside the burning wick, the magnesium is shielded from oxygen and cooled by liquid paraffin, but once the flame goes out, the ember ignites the magnesium dust. If you watch the ember, you will see tiny flecks of magnesium going off. One of them produces the heat necessary to relight the paraffin vapor, and the candle flame comes back to life!

How does dry ice work?

Dry ice is frozen carbon dioxide. A block of dry ice has a surface temperature of -109.3°F (-78.5°C). Dry ice also has the very nice feature of sublimation — as it melts, it turns directly into carbon dioxide gas rather than liquid. The super-cold temperature and the sublimation feature make dry ice great for refrigeration. For example, if you want to send something frozen across the country, you can pack it in dry ice. It will be frozen when it reaches its destination and there will be no messy liquid left over as you would have with normal ice.

Many people are familiar with liquid nitrogen, which boils at -320°F (-196°C). Liquid nitrogen is fairly messy and difficult to handle. So why is nitrogen a liquid while carbon dioxide is a solid? This difference is caused by the solid-liquid-gas features of nitrogen and carbon dioxide.

We are all familiar with the solid-liquid-gas behavior of water. We know that at sea level water freezes at 32°F (0°C) and boils at 212°F (100°C). Water behaves differently as you change the pressure, however. As you lower the pressure, the boiling point falls. If you lower the pressure enough, water will boil at room temperature. If you plot out the solid-liquid-gas behavior of a substance like water on a graph that shows both temperature and pressure, you create what's called a phase diagram for the substance. The phase diagram shows the temperatures and pressures at which a substance changes between solid, liquid, and gas.

At normal pressures, carbon dioxide moves straight between gas and solid. It is only at much higher pressures that you find liquid carbon dioxide. For example, a high-pressure tank of carbon dioxide or a carbon dioxide fire extinguisher contains liquid carbon dioxide.

Dry ice safety

If you ever have a chance to handle dry ice, you want to be sure to wear heavy gloves. The super-cold surface temperature can easily damage your skin if you touch it directly. For the same reason you should never try to taste or swallow dry ice.

Another important concern with dry ice is ventilation. You want to make sure the area is well ventilated. Carbon dioxide is heavier than air and it can concentrate in low areas or in enclosed spaces (like a car or a room where dry ice is sublimating). Normal air is 78% nitrogen, 21% oxygen, and only 0.035% carbon dioxide. If the concentration of carbon dioxide in the air rises above 5%, carbon dioxide can become toxic. Be sure to ventilate any area that contains dry ice.

14

To make dry ice, you start with a high-pressure container full of liquid carbon dioxide. When you release the liquid carbon dioxide from the tank, the expansion of the liquid and the high-speed evaporation of carbon dioxide gas cools the remainder of the liquid down to the freezing point, where it turns directly into a solid. If you have ever seen a carbon dioxide fire extinguisher in action, then you have seen this carbon dioxide snow form in the nozzle. You compress the carbon dioxide snow to create a block of dry ice.

How do mood rings work, and do they really tell my mood?

Mood rings were first seen as an extremely popular fad in the late 1970s, and they resurface regularly. The idea behind a mood ring is simple: Wear it on your finger and it will reflect the state of your emotions. The ring's stone should be dark blue if you're happy, and it supposedly turns black if you are anxious or stressed. While mood rings cannot reflect your mood with any real scientific accuracy, they actually are indicators of your body's involuntary physical reaction to your emotional state.

The stone in a mood ring is either a hollow glass shell filled with thermotropic (temperature-sensitive) liquid crystals, or a clear glass stone sitting on top of a thin sheet of liquid crystals. These liquid crystal molecules are very sensitive. They change position, or twist, according to changes in temperature. This change in molecular structure affects the wavelengths of light that are absorbed or reflected by the liquid crystals resulting in an apparent change in the color of the stone. For example, as the temperature increases, the liquid crystal molecules twist slightly in one direction. This twist causes the liquid crystal substance to absorb more of the red and green portions of the visible light, and reflect the blue part. Consequently, the stone appears dark blue. When the temperature decreases, the molecules begin to twist in the other direction and reflect a different portion of the spectrum.

The inside of the ring conducts heat from your finger to the liquid crystals in the "stone." The color green, which signifies "average" on the mood ring color scale, is calibrated to the surface temperature of a typical person, approximately 82°F (28°C). If your surface temperature varies far enough from the norm, then the liquid crystals in the stone alter enough to cause a change in the color reflected. And if you take off a mood ring, it will normally cool down and change to black unless the ambient temperature is very high.

Take a look at the mood ring colors listed below, and what "mood" they represent. The colors are listed according to the change in temperature they represent, with dark blue being the warmest and black being the coolest.

- Dark blue: Happy, romantic, or passionate
- Blue: Calm or relaxed
- Blue-green: Somewhat relaxed
- Green: Normal or average
- Amber: A little nervous or anxious
- Gray: Very nervous or anxious
- Black: Stressed, tense, or feeling harried

If you take a moment to think about the moods represented by the colors, you'll see a definite correlation between your body's surface temperature and the color of the liquid crystal. When you are in a passionate mood, your skin is usually flushed. This is a physical reaction to an emotion, causing the capillaries to open up on the surface of the skin and release heat. This brings about a slight change in the surface temperature of your body. When you are nervous or stressed, your skin may feel clammy. This physical reaction to your emotional state causes the capillaries to cut off blood flow to the skin, causing the surface temperature to drop.

⚙ **Web Links**

How Liquid Crystal Displays (LCDs) Work

How do scratch-and-sniff stickers work? What makes them last for years and years?

Scratch-and-sniff stickers sound like such a good idea, but they never really caught on except in children's books and the occasional perfume strip in a magazine.

Nonetheless, if you have kids you probably have a scratch-and-sniff book around the house. And even if the book is 20 years old, it still works! The reason the stickers last so long is because of the microencapsulation technology used to create them. The basic idea behind scratch-and-sniff is to encapsulate the aroma-generating chemical in gelatin or plastic spheres that are incredibly small — on the order of a few microns in diameter. When you scratch the sticker you rupture some of these spheres and release the smell. The smell is essentially held in millions of tiny bottles, and you break a few of the bottles every time you scratch the sticker. The tiny bottles preserve the fragrance for years.

The microencapsulation technology used in scratch-and-sniff stickers was first developed to create carbonless copy paper. In the copy paper, the top sheet of paper is coated with microcapsules containing colorless ink. When you write on the paper, the writing breaks the capsules and releases the ink. The ink mixes with a developer chemical on the next sheet of paper to create a dark color!

How does Alka-Seltzer work and why does it fizz? MB

The fizzing you see when you drop an Alka-Seltzer tablet into water is the same sort of fizzing that you see from baking powder. A baking powder reaction is caused by an acid reacting with baking soda (sodium bicarbonate). In school you probably tried an experiment in which you mixed baking soda with vinegar to see it foam; the reaction is what happens in baking powder.

If you look at the ingredients for Alka-Seltzer, you will find that it contains citric acid and sodium bicarbonate (baking soda). When you drop the tablet in water, the acid and the baking soda react and fizz. You can think of an Alka-Seltzer tablet as compressed baking powder with a little aspirin mixed in.

How does Pop Rocks candy work?

"Pop Rocks" is an extremely cool candy to some people, but to other people it is just plain weird and they won't touch the stuff. Whether you are for or against Pop Rocks, you have to admit that it is a technology candy — nothing in nature works like Pop Rocks does!

So how does the candy work? Pop Rocks is actually patented, so you can go read the patent and see exactly how they work. The patent number is 4,289,794.

Here's the basic idea. Hard candy (like a lollipop or a Jolly Rancher) is made from sugar, corn syrup, water, and flavoring. You heat the ingredients together to melt the sugar and boil the mixture to drive off all of the water. Then you let the temperature rise. What you are left with is a pure sugar syrup at approximately 300°F (149°C). When it cools, you have hard candy.

To make Pop Rocks, the hot sugar mixture is allowed to mix with carbon dioxide gas at about 600 pounds per square inch (PSI). The carbon dioxide gas forms tiny, 600 PSI bubbles in the candy. After the mixture cools you release the pressure and the candy shatters, but the remaining pieces still contain the high-pressure bubbles. You can see the bubbles if you use a magnifying glass to look at a piece of the candy. When you put the candy in your mouth, it melts (just like hard candy) and releases the bubbles with a loud POP! What you are hearing and feeling is the 600 PSI carbon dioxide gas being released from each bubble.

How do film companies color old black-and-white movies?

Most of the classic black and white movies have been "colorized," mainly so that they can be shown on television in color. The process used to add the color is fascinating but extremely tedious; someone has to work on the movie frame by frame, adding the colors one at a time to each part of the individual frame.

To speed up the process, the coloring is done on a computer

using a digital version of the film. The film is scanned into the computer and the coloring artist can view the movie one frame at a time on the computer's screen. The artist draws the outline for each color area and the computer fills it in. The original black-and-white film holds all of the brightness information, so the artist can paint large areas with a single color and let the original film handle the brightness gradients. This means that the artist might only have to add 10 or so actual colors to a scene.

To speed up the process even more, interpolation is common. From frame to frame in a single scene there is normally very little variation in the position of objects and actors. Therefore the artist might manually color every tenth frame and let the computer fill in the frames in between.

Light and Sound

3

⚙ Why is the sky blue? • Could I see a flashlight beam from the earth on the moon? • How do a zebra's stripes act as a camouflage since zebras do not live in a black-and-white environment? • What causes a sonic boom? • What is a light year? • How does a gun silencer work? • What is a decibel? • How do motion sensors work? • How does feedback work in a PA system to cause that howling sound? • How does an Indiglo watch work?

Why is the sky blue?

Here is something interesting to think about. When you look at the sky at night it is black, with the stars and the moon forming points of light on that black background. So why is it that, during the day, the sky does not remain black with the sun acting as another point of light? Why does the daytime sky turn a bright blue and the stars disappear?

The first thing to recognize is that the sun is an extremely bright source of light — much brighter than the moon. The second thing to recognize is that the atoms of nitrogen and oxygen in the atmosphere have an effect on the sunlight that passes through them. There is a physical phenomenon called "Rayleigh scattering" that causes light to scatter when it passes through particles that have a diameter 1/10 that of the wavelength (color) of the light. Sunlight is made up of all different colors of light, but because of the elements in the atmosphere, the color blue is scattered much more efficiently than other colors.

So when you look at the sky on a clear day, you see the sun as a bright disk. The blueness you see everywhere else is all of the atoms in the atmosphere scattering blue light toward you, but not scattering red light, yellow light, green light, or any of the other colors nearly as well.

Could I see a flashlight beam from the earth on the moon? TOP 20

This is a good question because it makes you think about how light works. When you turn on a flashlight, you are creating a source of photons. The photons leave the flashlight and immediately start to spread out in a cone-shaped beam. Provided that they don't hit anything, each individual photon travels through space forever. So it is not that the photons "run out of gas" on the way to the moon but that, by the time they reach the moon, the photons have spread out tremendously. As a result, so few photons hit your eye at any one time that you cannot detect the flashlight beam.

Whether or not you can see a flashlight beam projected from the earth to the moon depends on both the flashlight and on the size of your "eye." If the flashlight is a little penlight flashlight powered by a couple AA batteries, and if the eye is your naked eye, then the answer is, no—you cannot see the flashlight from the moon. The cone of a typical flashlight is gigantic by the time it reaches the moon, and the photons are spread too thinly for your naked eye to detect. If you were to use a much bigger flashlight (for example, an aircraft search light), or if you were to increase the size of your eye by using a telescope, then it is possible for you to detect the flashlight from the moon.

Another alternative would be to replace the flashlight with a small laser. The cone of divergence of a laser is extremely small compared to a flashlight. For example, one laser has a beam so tightly focused that, by the time the light reaches the moon, it has only diverged into a circle about half a mile (1 km) in diameter! You could easily see tightly focused laser light like that from the moon.

The other alternative is to increase the size of your eye by using a telescope. A telescope collects light over a large area with its lens or mirror. This is why people use large telescopes to detect the light from distant stars. Although the stars are very bright compared to a flashlight, they are also very far away; most stars are many light years away, and 1 light year equals 6 trillion miles or 10 trillion kilometers. By the time the star's light reaches the earth, therefore, the light is very dim.

Astronomers can see light from distant objects very clearly now with the Hubble Space Telescope. In fact, it has been said that the Hubble can detect the light from a match on Pluto!

○ Web links

How Lasers Work
How Light Works

○

How do a zebra's stripes act as a camouflage since zebras do not live a black-and-white environment?

To humans, a zebra's stripes stick out like a sore thumb, so it's hard to imagine that the stripes act as camouflage. Zoologists believe stripes offer zebras protection from predators in a couple different ways.

The first is as simple pattern camouflage, much like the type the military uses in its fatigue design. The wavy lines of a zebra blend in with the wavy lines of the tall grass around it. It doesn't matter that the zebra's stripes are black and white and the lines of the grass are yellow, brown, or green, because the zebra's main predator, the lion, is colorblind. The pattern of the camouflage is much more important than its color when hiding from these predators. If a zebra is standing still in matching surroundings, a lion may overlook it completely.

This benefit may help an individual zebra in some situations, but stripes are even better when zebras are in herds. Zebras usually travel in large groups, in which they stay very close to one another. Even with their camouflage pattern, it's highly unlikely that a large gathering of zebras would be able to escape a lion's notice, but their stripes help them use this large size to their advantage. When all the zebras keep together as a big group, the pattern of each zebra's stripes blends in with the stripes of the zebras around it. This is confusing to the lion, who sees a large, moving, striped mass instead of many individual zebras. The lion has trouble picking out any one zebra, and so it doesn't have a good plan of attack. It's hard for the lion to even recognize which way each zebra is moving. Imagine the difference in pursuing one animal and charging into an amorphous blob of animals moving every which way. The lion's inability to distinguish zebras also makes it more difficult for the lion to target and track weaker zebras in the herd.

So do zebra stripes confuse zebras as much as they confuse lions? Oddly enough, while making zebras indistinguishable to other animals, zebra stripes actually help zebras recognize one another.

Stripe patterns are like zebra fingerprints — every zebra has a slightly different arrangement. Zoologists believe this is how zebras distinguish who's who in a zebra herd. This patterning certainly has significant benefits. A zebra mare and her foal can keep track of each other in the large herd, for example, and a zebra can very quickly distinguish its own herd from another. This also helps human researchers, because it enables them to track particular zebras in the wild.

○ Web Links

How Safaris Work

What causes a sonic boom?

You can learn a lot about sonic booms by looking at the wakes boats leave in the water.

If you toss a pebble in a pond, little waves will form in concentric circles and propagate away from the point of impact. If a boat travels through the pond at 3-5 mph (4.8-8 kph), circular waves will propagate in the same way both ahead of and behind the boat, and the boat will travel through them.

If a boat travels faster than the waves can propagate through water, then the waves "can't get out of the way" of the boat fast enough, and they form a wake. A wake is a larger single wave that is formed out of all the little waves that would have propagated ahead of the boat but could not.

When an airplane travels through the air, it produces sound waves. If the plane is traveling slower than the speed of sound (the speed of sound varies, but 700 mph or 1,127 kph is typical through air), then sound waves can propagate ahead of the plane. If the plane breaks the sound barrier and flies faster than the speed of sound, it produces a sonic boom when it flies past. The boom is the "wake" of the plane's sound waves. All of the sound waves that would have normally propagated ahead of the plane are combined together. You hear nothing when the plane passes, but a moment later you hear the sonic boom.

Light and Sound

○

25

It is just like being on the shore of a smooth lake when a boat speeds past. There is no disturbance in the water as the boat comes by, but eventually a large wave from the wake rolls onto shore. When a plane flies past at supersonic speeds, the same thing happens, but instead of the large wave, you get a sonic boom.

How the Concorde Works

What is a light year?

A light year is a way of measuring distance. That definition doesn't make much sense because "light year" contains the word "year," which normally is a unit of time. Even so, light years measure distance.

You are used to measuring distances in either inches/feet/miles or centimeters/meters/kilometers, depending on where you live. You know the length of a foot or a meter and you are comfortable with these units because you use them — as well as miles or kilometers — every day. These are nice, human increments of distance.

When astronomers use their telescopes to look at stars, things are different. The distances are gigantic. For example, the closest star to earth (besides our sun) is something like 24,000,000,000,000 miles (39,000,000,000,000 kilometers) away. That's the closest star. Some stars are billions of times farther away. When you start talking about distances that far away, a mile or kilometer just isn't a practical unit to use because the numbers get too big. No one wants to write or talk about numbers that contain 20 digits.

So for really long distances, people use a unit called a light year to measure distance. Light travels at 186,000 miles per second (300,000 kilometers per second). Therefore, a light second is 186,000 miles (300,000 kilometers). A light year is the distance light could travel in a year, or:

186,000 miles/second x 60 seconds/minute x 60 minutes/hour x 24 hours/day x 365 days/year = 5,865,696,000,000 miles/year

A light year is 5,865,696,000,000 miles (9,460,800,000,000 kilometers). That's a long way!

Using a light year as a distance measurement has another advantage: It helps you determine age. Let's say that a star is 1 million light years away. The light from that star has traveled at the speed of light to reach us. Therefore it has taken the star's light 1 million years to get here, and the light we are seeing was created 1 million years ago. So the star we are seeing is really how the star looked 1 million years ago, not how it looks today. In the same way, our sun is approximately 8 light minutes away. If the sun were to suddenly explode right now, we wouldn't know about it for 8 minutes because that is how long it would take for the light of the explosion to get here.

A light nanosecond — the distance light can travel in a billionth of a second — is approximately 1 foot (30 centimeters). Radar uses this fact to measure how far away something like an airplane is. A radar antenna sends out a short radio pulse and then waits for it to echo off an airplane or other target. While it's waiting, it counts the number of nanoseconds that pass. Radio waves travel at the speed of light, so the number of nanoseconds divided by 2 tells the radar unit how far away the object is.

○ **Web Links**

How Time Works

How does a gun silencer work?

You might wonder how you can possibly take an explosive noise that can damage your hearing and turn it into a little "ffft" sound like you see in the movies. Gun silencers work on a very simple principle to silence guns.

Imagine a balloon. If you pop a balloon with a pin, it will make a loud noise. But if you untie the end of the balloon and let the air out slowly, it will make very little noise. That is the basic idea behind a gun silencer.

To fire a bullet from a gun, gunpowder is ignited behind the bullet. The gunpowder creates a high-pressure pulse of hot gas. The pressure of the gas forces the bullet down the barrel of the gun. When the bullet exits the end of the barrel, it is like uncorking a bottle. The pressure behind the bullet is immense, however — on the order of 3,000 pounds per square inch (PSI) — so the POP that the gun makes as it is uncorked is extremely loud.

A silencer screws on to the end of the barrel and is 20-30 times greater in volume than the barrel. With the silencer in place, the pressurized gas behind the bullet has a big space to expand into. So the pressure and temperature of the hot gas falls significantly. When the bullet finally exits through the hole in the silencer, the pressure being uncorked is much, much lower — perhaps 60 PSI. Therefore the sound the gun makes is much lower.

A bullet that travels at supersonic speeds cannot be silenced because the bullet creates its own little sonic boom as it travels. Many high-powered loads travel at supersonic speeds. The silencer can remove the "uncorking" sound, but not the sound of the bullet's flight.

What is a decibel?

The decibel (abbreviated dB) is the unit used to measure the intensity of a sound. The decibel scale is a little odd because the human ear is incredibly sensitive. Your ears can hear everything from your fingertip brushing lightly over your skin to a loud jet engine. In terms of power, the sound of the jet engine is about 1 trillion times more powerful than the smallest audible sound. That's a big difference!

On the decibel scale, the smallest audible sound (near total silence) is 0 dB. A sound 10 times more powerful is 10 dB. A sound 100 times more powerful than near total silence is 20 dB. A sound 1,000 times more powerful than near total silence is 30 dB. And so on. Here are some common sounds and their decibel ratings:

- Near total silence — 0 dB
- A whisper — 15 dB
- Normal conversation — 60 dB
- A lawnmower — 90 dB
- A car horn — 110 dB
- A rock concert or a jet engine — 120 dB

- A gunshot or a firecracker — 140 dB

All of these ratings are taken while standing near the sound. As you know from your own experience, distance affects the intensity of sound; if you are far away, the power is greatly diminished.

Any sound above 85 dB can cause hearing loss, and the loss is related both to the power of the sound as well as the length of exposure. For example, 8 hours of 90 dB sound can cause damage, and any exposure to 140 dB sound causes immediate damage (and causes actual pain). You will know that you are listening to an 85 dB sound if you have to raise your voice to be heard by somebody else.

How do motion sensors work?

There are many different ways to create a motion sensor. For example:

- Stores often contain a beam of light crossing the room near the door and a photosensor on the other side of the room. When a customer breaks the beam, the photosensor detects the change in the amount of light and rings a bell.

- Many grocery stores have automatic door openers that use a very simple form of radar to detect when someone passes near the door. The box above the door sends out a burst of microwave radio energy and waits for the reflected energy to bounce back. When a person moves into the field of microwave energy, the movement changes the amount of reflected energy or the time it takes for the reflection to arrive and the box opens the door. Because these devices use radar, they often set off radar detectors.

- The same thing can be done with ultrasonic sound waves, bouncing them off a target and waiting for the echo.

All of these are active sensors. They inject energy (light, microwaves, or sound) into the environment in order to detect a change of some sort.

The "motion sensing" feature on most lights (and security systems) is instead a passive system that detects infrared energy. These sensors are therefore known as PIR (Passive InfraRed) detectors or pyroelectric sensors. In order for a sensor to detect a person, the sensor must be sensitive to the temperature of that person. Humans, having a skin temperature of approximately 93°F (34°C), radiate infrared energy with a wavelength between 9 and 10 micrometers. Therefore the sensors are typically sensitive in the range of 8 to 12 micrometers.

The devices themselves are simple electronic components not unlike a photosensor. The infrared light bumps electrons off a substrate, and these electrons can be detected and amplified into a signal.

You have probably noticed that this light is sensitive to motion, but not to a person who is standing still. That's because the electronics package attached to the sensor is looking for a fairly rapid change in the amount of infrared energy it is seeing. When a person walks by, the amount of infrared energy in the field of view changes rapidly and is easily detected. You do not want the sensor detecting slower changes, like when a sidewalk cools off at night.

A motion-sensing light has a wide field of view because of the lens you can see covering the sensor. Infrared energy is a form of light, so you can focus and bend it with plastic lenses. But it is not like there is a 2-D array of sensors in there. Inside is a single sensor, or sometimes two sensors, looking for changes in infrared energy.

If you have a burglar alarm with motion sensors, you may have noticed that the motion sensors cannot "see" you when you are outside looking through a window. That is because glass is not very transparent to infrared energy.

How does feedback work in a PA system to cause that howling sound?

A simple PA system consists of a microphone, an amplifier, and one or more speakers. Whenever you have those three components you have the potential for feedback. Feedback occurs when the sound from the speakers makes it back into the microphone and is re-amplified and sent through the speakers again.

Imagine, for example, that you place a microphone in front of a speaker. Next you tap on the microphone. The sound of the tap goes through the amplifier, comes out the speaker, re-enters the microphone, and so on. This loop happens so quickly that it creates its own frequency, which you hear as a howling sound. The distance between the microphone and the speakers has a lot to do with the frequency of the howling, because that distance controls how quickly the sound can loop through the system.

Creating feedback on your PC:

You can try this concept on your computer if your setup includes speakers and a microphone:

1. In Windows, you need to enable the microphone and speakers by using the volume control. To do this, find the speaker icon (usually located in the bottom right-hand corner of the computer screen) and double-click on the speaker icon.

2. Make sure that in the dialog box the microphone and speakers are not muted and are at maximum volume. (If the microphone control is not visible, select it in Properties.)

3. If you have it set up right, you should be able to tap the microphone and hear it in the speakers.

4. Now place the microphone near the speakers and turn up the speaker volume until you hear the feedback. Try changing the distance between the microphone and speakers and see what effect these changes have. Be sure not to try this at 2 a.m. when other family members are sleeping, and also be sure to put the dog out…

If you are setting up a sound system and want to avoid feedback, a few general guidelines should help you avoid the problem:

- Make sure the speakers are in front of and pointing away from the microphone. If the speakers are behind the microphone, feedback is nearly guaranteed.
- Use a unidirectional microphone.
- Place the microphone close to the person who is speaking/performing.
- If you have access to an equalizer, dampen the frequencies where feedback is occurring.

How does an Indiglo watch work?

The basic technology behind the Indiglo watch is called electro-*luminescence*. Electroluminescence is the conversion of electricity directly into light. This is not how an incandescent bulb works. In an incandescent bulb the electricity produces heat and the heat produces light. In the same way that you can heat a horse-shoe to "white hot" so that it glows brilliantly, a bulb's filament is "white hot." Electroluminescence is much more efficient because it converts the electricity directly to light.

The most common example of electroluninescence you'll see regularly is a neon light. In a neon light, high voltage energizes the electrons in neon atoms and, when the electrons de-energize themselves, they emit photons.

In an Indiglo watch, a very thin panel uses high voltage to ener-gize phosphor atoms that produce light. The panel itself is extremely simple. As described in the Timex patent, you take a thin glass or plastic layer, coat it with a clear conductor, coat that with a very thin layer of phosphor, coat the phosphor with a thin plastic, and then add another electrode. Essentially what you have is two conductors (a capacitor) with phosphor in between. When you apply 100 to 200 volts of alternating current (AC) to the conductors, the phosphor energizes and begins emitting pho-tons.

Creating the high voltage can be a problem in a wristwatch because the watch has only a small 1.5-volt battery. To produce the 100-200 volts, a 1:100 transformer is used. By charging the primary coil of the transformer with a transistor that is switching on and off, the secondary coil rises to 150 volts or so.

But I Thought GIF was Peanut Butter . . .

4

⚙ Does adding more RAM to my computer make it faster? • What do the different Web page extensions mean (html, htm, asp, and so on)? • What is Linux? • Is it better to turn my computer off when I am not using it, or to leave it on all the time? • What is a T1 line? • Why are there so many different image formats on the Web? For example, what is the difference between a GIF and a JPG image? • What is the fastest computer in the world? • What is the Year 2038 Problem?

Does adding RAM to my computer make it faster?

Up to a point, adding random access memory (RAM) will normally cause your computer to run faster on certain types of operations. RAM is important because of an operating system component called the virtual memory manager (VMM).

When you run a program such as a word processor or an Internet browser, the microprocessor in your computer pulls the executable file (EXE) for the program off the hard disk and loads it into RAM. In the case of a big program like Microsoft Word or Excel, the EXE consumes about 5 megabytes of RAM. The microprocessor also pulls in a number of dynamic link libraries (DLLs) — shared pieces of code used by multiple applications. The DLLs might total 20-30 megabytes. Then the microprocessor loads in the data files that you want to look at, which might add several more megabytes of RAM if you are looking at several documents or browsing a page with a lot of graphics. So a normal application needs from 10-30 megabytes of RAM space to run. The average machine at any given time might have the following applications running:

- A word processor
- A spreadsheet
- A DOS prompt
- An email program
- A drawing program
- Three or four browser windows
- A fax program
- A Telnet session

Besides all of those applications, the operating system itself is taking up a good bit of space. Those programs together might need 100-150 megabytes of RAM, but the computer might have only 64 megabytes of RAM installed.

The extra space needed is created by the VMM. The VMM looks at RAM and finds sections that are not currently needed. It puts these sections of RAM in a place called the swap file on the hard disk. If the email program is running but you have not looked at email for 45 minutes, the VMM moves all the bytes making up the email program's EXE, DLLs, and data out to the hard disk. That is called swapping out the program. The next time you click on the email program, the VMM will swap in all of its bytes from the hard disk, and probably in the process swap something else out. Because the hard disk is slow relative to RAM, the act of swapping things in and out causes a noticeable delay.

If you have a very small amount of RAM (say 16 megabytes), then the VMM is always swapping things in and out to get anything done. In that case, your computer feels like it is crawling. As you add more RAM, you get to a point where you only notice the swapping when you load a new program or change windows. If you were to put 256 megabytes of RAM in your computer, the VMM would have plenty of room and you would never see it swapping anything. Past that point, however, adding more RAM will have no effect on the computer's speed.

Some applications need tons of RAM to do their job. If you run them on a machine with too little RAM, they swap constantly and run very slowly. You can get a huge speed boost by adding enough RAM to eliminate the swapping. Programs like these may run 10-50 times faster once they have enough RAM.

⚙ Web Links

How Microprocessors Work

What do the different Web page extensions mean (html, htm, asp, and so on)?

The seven most common extensions that you see on the end of URLs are:

- .htm
- .html
- .shtml
- .nsf
- .asp

⚙

- .pl
- .cgi

When the Web started, it ran almost exclusively on UNIX machines and all pages were static. The standard file extension was .html. When people started using PCs running DOS or Windows as Web servers, however, the four letters in .html were problematic. PCs followed an 8.3 naming convention that allowed only three letters in the extension. So the world made room for two standard extensions: .html and .htm.

Pages tagged with .htm or .html are static; the file is lifted off the server's disk and sent verbatim to the client.

Pages tagged with .shtml, however, reveal that server side includes (SSI) are being used on the server. With SSI a page can contain tags to indicate that another file should be inserted in place of the tag in the existing page. So a page is lifted off the server's disk and the server makes all the substitutions indicated. It then sends the final page to the client. This approach enables you to easily change elements like headers and footers on pages across an entire site.

You will see the .nsf extension if a site is powered by Lotus's Domino server. This server takes standard Lotus Notes databases and dynamically converts the content of those databases into Web pages when the page is requested. The site's pages can change throughout the day depending on the content of the databases.

The extension .asp indicates that the page uses Active Server Pages (ASP), which is a Microsoft technology. A Web page using ASP can contain Visual Basic code that the server executes when it lifts a page off the disk. This code can do just about anything: read databases, run other programs, custom format pages based on the user's ID, and so on.

The .pl extension stands for PERL, which is a scripting language. Pages with the .pl extension contain nothing but PERL script, and the script builds the page on the fly.

The extension .cgi also indicates that a page contains code executed by the server, but the type of code could be just about anything. On HowStuffWorks.com, for example, C++ code is compiled to create .cgi files.

Web Links

How CGI Scripting Works

What is Linux?

Every desktop computer uses an operating system. The most popular operating systems in use today are:

- Windows
- The Mac OS
- UNIX

Linux is a version of the UNIX operating system that has become very popular over the last several years.

Operating systems are computer programs. An operating system is the first piece of software that the computer executes when you turn on the machine. The operating system loads itself into memory and begins managing the resources available on the computer. It then provides those resources to other applications that the user wants to execute.

The operating system is the overall housekeeper of your computer and controls such functions as scheduling the various tasks the computer must do, controlling and distributing the system's RAM, creating and maintaining the directories and files on the hard disk, and controlling all the data that moves between the computer and the network. It also manages various computer components like the keyboard, mouse, video display, and printers and maintains the security of the information in the computer's files by controlling which users can access the computer.

An operating system normally also provides the default user interface for the system. The standard "look" of Windows includes such features as the Start button and the task bar. The Mac OS provides a completely different look and feel for Macintosh computers.

Linux is as much a phenomenon as it is an operating system. The Linux kernel, created by Linus Torvalds, was made available to the world for free. Torvalds then invited others to add to the kernel, provided that they kept their contributions free as well. Thousands of programmers began working to enhance Linux and the operating system grew rapidly. Because Linux is free and runs on PC platforms, it quickly gained a sizeable audience among hard-core developers. Linux has a dedicated following and appeals to several different kinds of people:

- People who already know UNIX and want to run it on PC-type hardware
- People who want to experiment with operating system principles
- People who need or want a great deal of control over their operating system

In general, Linux is harder to manage than an operating system like Windows, but it offers more flexibility and configuration options.

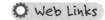 Web Links

How Operating Systems Work

Is it better to turn my computer off when I am not using it, or to leave it on all the time?

This is one of those questions that has no single right answer; the correct answer for you depends on how you use your computer. Here are some thoughts to keep in mind, and then you can decide which scenario applies to you.

Let's start with three situations that force you to leave your computer on 24 hours a day. They are:

- You are on a network, and the network administrators back up files and/or upgrade software over the network at night. In this situation, if you want your machine backed up or upgraded, then you need to leave it on all the time.
- You are using your machine as some sort of server. If your machine acts as a file server, print server, or Web server on a LAN or the Internet, then you need to leave it on all the time.
- If you are running something like SETI@home and you want to produce as many result sets as possible, you need to leave your machine on all the time.

If you do not fall into any of those categories, then you have a choice about whether or not to leave your machine on.

One reason why you might want to turn it off is economic. A typical PC and monitor consume approximately 300 watts. Let's assume that you use your PC for 4 hours every day, so the other 20 hours it is on would be wasted energy. To calculate kilowatt-hours, multiply watts times hours and divide by 1,000. So, if electricity costs 10 cents per kilowatt-hour in your area, then those 20 hours represent 60 cents a day. Sixty cents a day adds up to $219 per year — a pretty big chunk of money.

It's possible to use the energy-saving features built into modern machines and cut that expense in half. For example, you can have the monitor and hard disk power down automatically when not in use.

You can decide which approach works best for you based on how you use your computer.

What is a T1 line?

If your office has a T1 line, it means that the phone company has brought a fiber optic line into your office. (A T1 line might also come in on copper.) A T1 line can carry 24 digitized voice channels, or it can carry data at a rate of 1.544 megabits per second. If the T1 line is being used for telephone conversations, it plugs into the office's phone system. If it is carrying data, it plugs into the network's router.

A T1 line can carry about 192,000 bytes per second — roughly 60 times more data than a normal residential modem. It is also extremely reliable — much more reliable than an analog modem. Depending on how the users are using the line, a T1 line can generally handle quite a few people. For general Web browsing, hundreds of users are easily able to share a T1 line comfortably. If all of those users are downloading MP3 files or video files simultaneously it would be a problem, but this type of simultaneous use isn't extremely common.

A T1 line might cost between $1,000 and $1,500 per month depending on who provides it and where it goes. One end of the T1 line needs to be connected to an Internet service provider (ISP), and the total cost is a combination of the fee the phone company charges and the fee the ISP charges.

A large company needs more than a T1 line. The following table shows some of the common line designations:

DS0	64 kilobits per second
ISDN	Two DS0 lines plus signaling (16 kilobits per second), or 144 kilobits per second
T1	1.544 megabits per second (24 DS0 lines)
T3	43.232 megabits per second (28 T1s)
OC3	155 megabits per second (100 T1s)
OC12	622 megabits per second (4 OC3s)
OC48	2.5 gigabits per second (4 OC12s)
OC192	9.6 gigabits per second (4 OC48s)

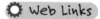

How Web Servers and the Internet Work

Why are there so many different image formats on the Web? For example, what is the difference between a GIF and a JPG image?

It certainly is true that there are lots of different image formats on the Web; at HowStuffWorks.com we use five different image formats:

- GIF files
- JPG files
- Animated GIF files
- MPG files
- Shockwave files

The two most common by far are GIF and JPG files. Both of these formats encode static bitmap images.

In a bitmap image, the image file has to define the exact color of every pixel in the image. For example, imagine a typical bitmap on the Web that is 400 × 400 pixels. To define this image, you would need 24 bits per pixel for 160,000 pixels, or 480,000 bytes. That would be a huge image file, so both the GIF and JPG formats compress the image in different ways.

In a GIF image, the number of colors is reduced to 256, and then "runs" of same-color pixels are encoded in a color+numberOfPixels format. For example, if there are 100 pixels on a line with the color 41, the image file stores the color (41) and the length of the run (100). This makes a GIF file great for storing drawings that have lots of same-color pixels.

A JPG file uses a more complex technique to compress images, like photographs, where the color of every pixel is different. In the process, the colors of individual pixels can change slightly, but the overall look of the picture is preserved.

An animated GIF is a sequence of GIF files all bonded together and displayed one after the other. With enough panes, you can get very realistic animations. However, the size of the file is the sum of the GIF files used to create the sequence, and that number can add up quickly.

An MPG file uses a complex algorithm like a JPG file does: It tries to eliminate repetition between frames to significantly compress video information. In addition, it allows a soundtrack, which animated GIFs do not. Because a typical sequence has hundreds or thousands of frames, file sizes can still get quite large.

Shockwave provides a vector-based animation capability. Instead of specifying the color of every pixel, a Shockwave file specifies the coordinates of shapes — things like lines, rectangles, and circles — as well as the color of each shape. Shockwave files allow animation and sound and can be extremely small. The images are also scalable; because they are vector-based, you can enlarge the image and it will still look great.

Each file format has a set of features and advantages that make it the best choice in a given situation. That is why there are so many image formats.

What is the fastest computer in the world?

To put things in perspective, let's start with the computer sitting on your desk — the computer you use on a day-to-day basis to browse the Internet, handle spreadsheets, and create documents. Most people have something like a Pentium computer running

Windows, or a Macintosh computer using the Mac OS. A computer like this can execute approximately 100 million instructions per second. Your particular machine might be twice that fast or half that fast, but that's the ballpark.

The fastest computer in the world is much faster than that, and it is sitting right on top of your shoulders! Although we take our brains for granted, they are astonishing computing devices and they are the fastest processors available right now. Let me give you an example. Your desktop computer is just starting to get to the point where it can "understand" speech and take dictation, translating spoken words into written words. It can only understand one speaker, and that speaker has to train it for about 20 minutes, and the dictation software will still make a lot of mistakes. So 100 million instructions per second can barely handle dictation.

Your brain, on the other hand, can understand any number of speakers, and it will make zero mistakes. It may even be able to understand multiple languages! And the speech processing portion of your brain is just one small part of the whole package; your brain can also process complex visual images, control your entire body, understand conceptual problems and create new ideas, and perform many more amazing tasks. Your brain is made up of about 1 trillion cells with 100 trillion connections between those cells. We might take a rough estimate and say that your brain is handling 10 quadrillion instructions per second, but it really is hard to say.

It will be several decades before we develop a computer that can perform at the same rate as your brain. It will be decades before it is small enough to fit in a brain-sized package.

What is the Year 2038 Problem?

Most programs written in the C programming language were relatively immune to the Y2K problem, but suffer instead from the Year 2038 Problem. The problem arises because most C programs use a library of routines called the standard time library. This library establishes a standard 4-byte format for the storage of time values, and also provides a number of functions for converting, displaying, and calculating time values.

The standard 4-byte format assumes that the beginning of time is Jan. 1, 1970 at 12:00:00 a.m. This value is 0. Any time/date value is expressed as the number of seconds following that zero value. So the value 919642718 is 919,642,718 seconds past 12:00:00 a.m. on Jan. 1, 1970, which is Sunday Feb. 21, 1999 at 16:18:38 Pacific time in the U.S. This format is convenient because you can subtract any two values and get a number of seconds that is the time difference between them. Then you can use other functions in the library to determine how many minutes/hours/days/months/years have passed between the two times.

A signed 4-byte integer has a maximum value of 2,147,483,647, which is the crux of the Year 2038 Problem. The maximum value of time possible before this value rolls over to a negative (invalid) value is 2,147,483,647, which translates into January 19, 2038. On this date, any C programs that use the standard time library will start to have problems with date calculations.

Fortunately, this problem is somewhat easier to fix than the Y2K problem was to fix on mainframes. Well-written programs can simply be recompiled with a new version of the library that uses, for example, 8-byte values for the storage format. This is possible because the library encapsulates the whole time activity with its own time types and functions (unlike most mainframe programs, which did not standardize their date formats or calculations).

On The Go

⚙ What is the difference between gasoline, kerosene, diesel fuel, and other forms of oil? • What is the difference between a turbocharger and a supercharger on a car's engine? • How does a Jake Brake work on a big rig? • What does the "weight" mean on a can of motor oil? • How does a car's rearview mirror work when it's set on the upward, glare-resistant setting? • How does a traffic light detect that a car has pulled up and is waiting for the light to change? • How does a gasoline pump at a filling station know when the tank is full? • Why do some engines use a dry sump oil system? • If diesel engines are more efficient and use cheaper fuel than traditional gasoline engines, why don't all cars use diesel engines? • Is it true that a diesel engine can operate under water while a gasoline engine cannot? • When I go to the gas station, I always have a choice of three different octanes: 87, 89, and 93. What is the difference, and what does "unleaded" mean?

45

What is the difference between gasoline, kerosene, diesel fuel, and other forms of oil?

The "crude oil" pumped out of the ground is a black liquid called petroleum. This liquid contains aliphatic hydrocarbons, or hydrocarbons composed of nothing but hydrogen and carbon. The carbon atoms link together in chains of different lengths.

It turns out that hydrocarbon molecules of different lengths have different properties and behaviors. For example, a chain with just one carbon atom in it (CH_4) is the lightest chain, known as methane. Methane is a gas so light, in fact, that it floats like helium. As the chains get longer, they get heavier. The first four chains (CH_4, C_2H_6, C_3H_8, and C_4H_{10} or methane, ethane, propane, and butane) are all gases that boil at -161°F, -88°F, -46°F, and -1°F (-107°C, -71°C, -31°C, and 17°C), respectively. The chains up through $C_{18}H_{32}$ are all liquids and the chains above C_{19} are solids at room temperature.

The different chain lengths all have progressively higher boiling points, so they can all be separated by distillation. This is what happens in an oil refinery — crude oil is heated and the different chains are pulled out by their vaporization temperatures.

The chain lengths in the C_5, C_6, and C_7 range are all very light, easily vaporized clear liquids called *naphthas*. They are used as solvents; dry cleaning fluids are made from these liquids, as are paint solvents and other quick-drying products.

The chain lengths from C_7H_{16} through $C_{11}H_{24}$ are blended together and used for gasoline. All of them vaporize at temperatures below the boiling point of water.

Kerosene falls in the C_{12} to C_{15} range, followed by diesel fuel and heavier fuel oils (like heating oil for houses).

Lubricating oils no longer vaporize in any way at normal temperatures; for example, engine oil can run all day at 250°F (121°C) without vaporizing at all. Oils go from very light (like 3-in-1 oil) through various thicknesses of motor oil through very thick gear oils and then semi-solid greases.

Chains above the C_{20} range form solids, starting with paraffin wax; then comes tar, and finally asphaltic bitumen used to make asphalt roads.

All of these different substances come from crude oil. The only difference is the length of the carbon chains.

What is the difference between a turbocharger and a supercharger on a car's engine?

Both turbochargers and superchargers are called forced induction systems. They both compress the air flowing into the engine. The advantage of compressing the air is that it lets the engine stuff more air into a cylinder. More air means that more fuel can be stuffed in, as well. Therefore, you get more power from each explosion in each cylinder. A turbocharged or supercharged engine produces more power overall than the same engine without the charging.

The typical boost provided by either a turbocharger or a supercharger is 6-8 pounds per square inch (PSI). Because normal atmospheric pressure is 14.7 PSI at sea level, you can see that you get about 50% more air into the engine by using a turbocharger or supercharger. Therefore, you would expect to get 50% more power. Because these engines aren't perfectly efficient, however, you might get a 30%-40% improvement instead.

The key difference between a turbocharger and a supercharger is in their power supplies. Something has to supply the power to run the air compressor. In a supercharger, a belt connects directly to the engine. It gets its power the same way that the water pump or alternator does. A turbocharger gets its power from the exhaust stream. The exhaust runs through a turbine, which in turn spins the compressor.

There are tradeoffs with both systems. In theory, a turbocharger is more efficient because it is using the "wasted" energy in the exhaust stream for its power source. On the other hand, a

On The Go

turbocharger causes some amount of back pressure in the exhaust system, and it also tends to provide less boost until the engine is running at higher revolutions per minute (RPMs).

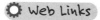 Web Links

How a Car Engine Works

How does a Jake Brake work on a big rig?

Many large diesel trucks (and even some larger RVs) are equipped with "Jake Brakes," also known as compression release engine braking systems. They are called Jake Brakes because Jacobs Vehicle Systems is the original maker of this braking system.

The basic idea behind a Jake Brake is to use the engine to provide braking power. If you own a stick-shift car and have ever downshifted to provide braking, you understand part of the idea. When you brake a car by downshifting, you are using engine vacuum to slow down the car.

A Jake Brake goes a step further, and actually turns the engine into an air compressor to provide a great deal more braking power. Engines go through a compression stroke, and this compressing of air in the cylinder requires power. If the engine's drive shaft is turning the engine to brake the truck, the power used to compress the air is braking power. That power is stored in the cylinder, however, so if you let it, the compressed air simply pushes the piston back down. Therefore you don't really get any braking from the compression stroke on an unmodified engine.

A Jake Brake modifies the timing on the exhaust valves so that, when braking is desired, the exhaust valves open right as the piston reaches the top of the compression stroke.

The energy gathered in the compressed air is released, so the compression stroke actually provides braking power.

The main advantage of a Jake Brake is that it saves wear on the normal brakes. This feature is especially important on long downhill stretches.

48

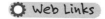 Web Links

How a Car Engine Works

What does the "weight" mean on a can of motor oil?

The weight on a can of oil actually reflects the oil's *viscosity*. Viscosity is a measurement of a fluid's ability to flow — the higher the viscosity, the better the fluid's ability to flow. Water, for example, has low viscosity. Honey, on the other hand, has high viscosity. The standard way to measure viscosity is to place the liquid in a cup that has a small hole in the bottom. Low-viscosity liquids will flow through the hole very quickly compared to high-viscosity liquids.

Temperature can have a big effect on viscosity. Liquids thin out when they are hot. If you heat up honey or cooking oil, for example, it gets much thinner.

In a car engine, you need oil that is as thin as possible when it is cold so that the engine starts more easily on a winter day. Unfortunately, very thin oil will thin out even more, and consequently provide no protection once the engine reaches its operating temperature.

To solve this problem, engine oils have two viscosities. For example, 5W-30 weight motor oil has a viscosity rating of 5 when it is cold. When it gets hot, it has a viscosity rating that 30-weight oil would have at the same temperature.

To give oil the ability to change viscosity as it heats up, the manufacturer adds a polymer to the oil. When it is cold, the polymer has no effect on the oil. As it heats up, the polymer activates and uncoils into longer thread-like molecules that increase the viscosity. The oil has, essentially, two personalities depending on its temperature!

○ Web Links

How Oil Refining Works

On The Go

How does a car's rearview mirror work when it's set on the upward, glare-resistant setting?

If you stand in front of a normal window in your house at night (inside the house, with the indoor lights on), you will be able to see your reflection in the glass quite clearly although the reflection is rather dim compared to a reflection in a mirror. A car's mirror takes advantage of the same effect. The mirror is not ground flat — the front glass surface is at an angle to the back (mirrored) surface. So if you look at this mirror out of its casing, it would be wedge shaped with the thicker edge at the top. When you "flip" the mirror, the back-mirrored surface actually points toward the dark ceiling, so you don't see that image. What you see instead is the image reflecting off the front of the glass, and this is much dimmer than the pure reflected image so it does not hurt your eyes.

To prove this is happening, take a flashlight with you one night and play with your mirror (while the car is parked in your garage, preferably). Flip the mirror and shine the light at the ceiling (or the floor); the fully reflected image will overwhelm the front-surface reflection so that you can see the ceiling.

How does a traffic light detect that a car has pulled up and is waiting for the light to change?

TOP
20

There is something exotic about the traffic lights that "know" you are there — the instant you pull up, they change.

These lights may detect when a car arrives at an intersection, when too many cars are stacked at an intersection (to control the length of the light), or when cars have entered a turn lane (in order to activate the arrow light).

There are all sorts of technologies for detecting cars — everything from lasers to rubber hoses filled with air. By far the most common technique is the inductive loop. An inductive loop is simply a coil of wire embedded in the road's surface. To install the loop, workers lay the asphalt and then use a saw to cut a groove in the asphalt. The wire is laid in the groove and sealed with a rubbery compound. You can often see these big rectangular loops cut in the pavement because the compound is obvious.

Inductive loops work by detecting a change of inductance. The coil in the road is an inductor. The inductance of an inductor is controlled by two factors:

- The number of coils
- The material that the coils are wrapped around (the core)

Iron in the core of an inductor gives the inductor much more inductance than air or any other non-magnetic core would provide. Devices exist that can measure the inductance of a coil, and the standard unit of measure is called a henry.

Let's say you take a coil of wire perhaps 5 feet in diameter, containing five or six loops of wire. You cut some grooves in a road and place the coil in the grooves. You attach an inductance meter to the coil and measure the inductance of the coil. Now you park a car over the coil and check the inductance again. The inductance will be much larger because of the large steel object positioned in the loop's magnetic field. The car parked over the coil is acting like the core of the inductor and its presence changes the inductance of the coil.

A traffic light sensor uses the loop in that same way. It constantly tests the inductance of the loop in the road and, when the inductance rises, it knows a car is waiting.

How does a gasoline pump at a filling station know when the tank is full?

It's actually a purely mechanical device, but an ingenious one.

Near the tip of the nozzle is a small hole, and a small pipe leads back from the hole into the handle. Suction is applied to this pipe using a venturi. When the tank is not full, air is drawn through the hole by the vacuum and the air flows easily. When gasoline in the tank rises high enough to block the hole, a mechanical linkage in the handle senses the change in suction and flips the nozzle off. Here's a way to think about it: Imagine a small pipe with suction being applied at one end and air flowing through the pipe easily. If you stick the free end of the pipe in a glass of water, much more suction is needed, so a vacuum develops in the middle of the pipe. That vacuum can be used to flip a lever that cuts off the nozzle.

The next time you fill up your tank, look for this hole either on the inside or the outside of the tip of the nozzle.

Why do some engines use a MB dry sump oil system?

Most production cars have a wet sump oil system. The sump is the area below the crankshaft in an engine. In a wet sump, the oil that you put into the engine is stored beneath the crankshaft in the oil pan. This pan has to be large and deep enough to hold 4-6 quarts (liters) of oil; think about two 3-liter bottles of soda and you can see that this storage area is pretty big.

In a wet sump, the oil pump sucks oil from the bottom of the oil pan through a tube, and then pumps it to the rest of the engine.

In a dry sump, extra oil is stored in a tank outside the engine rather than in the oil pan. There are at least two oil pumps in a dry sump: one pulls oil from the sump and sends it to the tank, and the other takes oil from the tank and sends it to lubricate the engine. The minimum amount of oil possible remains in the engine.

Dry sump systems have several important advantages over wet sumps:

- Because a dry sump does not need to have an oil pan big enough to hold the oil under the engine, the main mass of the engine can be placed lower in the vehicle. This helps lower the center of gravity and can also help aerodynamics (by allowing the height of the hood to be lower).

- The oil capacity of a dry sump can be as big as you want. The tank holding the oil can be placed anywhere on the vehicle.
- In a wet sump, turning, braking, and acceleration can cause the oil to pool on one side of the engine. This sloshing can dip the crankshaft into the oil as it turns or uncover the pump's pick-up tube.
- Excess oil around the crankshaft in a wet sump can get on the shaft and cut horsepower.

The disadvantage of the dry sump is the increased weight, complexity, and cost from the extra pump and the tank — a small price to pay for such big benefits.

If diesel engines are more efficient and use cheaper fuel than traditional gasoline engines, why don't all cars use diesel engines?

Diesel engines have never really caught on in passenger cars. During the late 1970s, diesel engines in passenger cars did see a surge in sales because of the OPEC oil embargo (more than half a million were sold in the U.S.), but that is the only significant penetration diesel engines have made in the marketplace. Even though diesel engines are more efficient than traditional gasoline engines, several historical problems have held diesel engines back:

- Diesel engines, because they have much higher compression ratios (20:1 for a typical diesel vs. 8:1 for a typical gasoline engine), tend to be heavier than an equivalent gasoline engine.
- Diesel engines also tend to be more expensive.
- Diesel engines, because of the weight and compression ratio, tend to have lower maximum revolutions per minute (RPM) ranges than gasoline engines. This makes diesel engines high torque rather than high horsepower, and that tends to make diesel cars slow in terms of acceleration.

- Diesel engines must be fuel injected, and in the past, fuel injection was expensive.
- Diesel engines tend to produce more smoke and "smell funny."
- Diesel engines are harder to start in cold weather, and if they contain glow plugs, diesel engines can require you to wait before starting the engine so that the glow plugs can heat up.
- Diesel engines are noisier and tend to vibrate.
- Diesel fuel is less readily available than gasoline.

One or two of these disadvantages would be OK, but a group of disadvantages this large is a big deterrent for lots of people.

The two things working in favor of diesel engines are better fuel economy and longer engine life. Both of these advantages mean that over the life of the engine you will tend to save money with a diesel. To calculate your true savings, however, you also have to take into account the initial high cost of the engine. You need to own and operate a diesel engine for a fairly long time before the fuel economy overcomes the increased purchase price of the engine. The equation works great in a big diesel tractor-trailer rig that is running 400 miles every day, but it is not nearly so beneficial in a passenger car.

As mentioned, the list above contains historical disadvantages for diesel engines. New diesel engine designs using advanced computer control are eliminating several of these disadvantages; smoke, noise, vibration, and cost are all declining. In the future, we are likely to see a lot more diesel engines on the road.

○ **Web Links**

How Diesel Engines Work

Is it true that a diesel engine can operate under water while a gasoline engine cannot?

You sometimes see military vehicles running in "extreme" conditions, and these conditions can include deep submersion during river fording. The air intake usually provides the first problem for operating in water; as soon as you submerge the air intake, the engine can no longer get air and it will stop running.

You can get around the air intake problem by adding a snorkel. Military Humvees often have a snorkel attached, and it allows them to submerge in up to 5 feet (1.6 meters) of water. In order to handle submersion like this, the engine and the rest of the vehicle must be waterproofed.

To waterproof any off-road vehicle, many different things must be thought about. For example:

- Any electrical devices must be sealed, including, the instruments, engine control computers, motors (for fans, windshield wipers, and so on), lights, and battery.

- Any venting for elements like the crankcase and differential must be sealed (or vented at the same level as the snorkel).

- The fuel tank must be sealed and vented appropriately.

- Any chamber or crevice that can fill with water must have a drain.

Assuming that all of these other things have been taken care of, and assuming that the engine is waterproofed, then the vehicle can run underwater.

In general, it is easier to waterproof a diesel engine than it is to waterproof a gasoline engine because of the spark plugs and ignition system in a gasoline engine. These components run at high voltage, and sealing them is harder (but not impossible). A diesel engine, on the other hand, has no ignition system. If the diesel engine has a mechanical fuel pump for the injectors and a mechanical transmission, there are also no engine control electronics to worry about. These features can make a diesel engine relatively easy to waterproof. That's why most military vehicles that ford rivers or run submerged have diesel engines.

⚙ **Web Links**

How a Car Engine Works
How a Diesel Engine Works

On The Go

⚙

When I go to the gas station, I always have a choice of three different octanes: 87, 89, and 93. What is the difference, and what does "unleaded" mean?

Almost all cars use 4-stroke gasoline engines. One of the strokes is the compression stroke, where the engine compresses a cylinder-full of air and gas into a much smaller volume before igniting it with a spark plug. The amount of compression is called the compression ratio of the engine. A typical engine might have a compression ratio of 8-to-1. This means that the volume in the cylinder when the piston is at the bottom of its stroke (before the air and fuel are compressed) is eight times the volume of the cylinder when the piston is at the top of its stroke (the air and fuel are fully compressed).

The octane rating of gasoline tells you the amount that the fuel can be compressed before it spontaneously ignites. When gas ignites by compression rather than because of the spark from the spark plug, it causes knocking in the engine. Knocking can damage an engine, so it is not something you want to have happening. Lower octane gas (like "regular" 87 octane gasoline) can handle the least amount of compression before igniting.

The compression ratio of your engine determines the octane rating of the gas you must use in the car. One way to increase the horsepower of an engine of a given displacement is to increase its compression ratio. So a "high performance engine" has a higher compression ratio and requires higher-octane fuel. The disadvantage is that the gasoline for your engine costs more.

The name "octane" comes from the following fact. When you take crude oil and "crack" it in a refinery, you end up getting hydrocarbon chains of different lengths. These different chain lengths can then be separated from each other and blended to form different fuels. For example, you may have heard of methane, propane, and butane. All three are hydrocarbons. Methane has just a single carbon atom. Propane has three carbon atoms chained together. Butane has four carbon atoms chained together. Pentane has five, hexane has six, heptane has seven, and octane has eight carbons chained together.

Heptane handles compression very poorly; compress it just a little and it ignites spontaneously. Octane handles compression very well; you can compress it a lot and nothing happens. "Regular" 87-octane gasoline contains 87 percent octane and 13 percent heptane (or some other combination of fuels that has the same performance of the 87/13 combination of octane and heptane). It spontaneously ignites at a given compression level, and can only be used in engines that do not exceed that compression ratio.

During World War I, it was discovered that a chemical called tetraethyl lead could be added to gasoline and significantly improve its octane rating. Cheaper grades of gasoline could be made usable by adding this chemical. This led to the widespread use of "ethyl" or "leaded" gasoline. Unfortunately, lead clogs a catalytic converter and renders it inoperable within minutes. When lead was banned, gasoline got more expensive because refineries could no longer boost the octane ratings of cheaper grades. Airplanes are still allowed to use leaded gasoline, and octane ratings of 115 are commonly used in super-high performance piston airplane engines.

◯ Web Links

How a Car Engine Works

Power Up!

⚙ Is it possible to generate electricity directly from heat? • Is flour flammable? I heard if you were to burn flour, it would explode. If so, why does this happen? • How does an oxygen canister work? • What home appliances use the most power? • Do commercial jets have locks on the doors and ignition keys? • My lawn mower doesn't have a battery like a car does, so where does the electricity to spark the spark plug come from? • How does a laser speed gun work to measure a car's speed compared to normal police radar? • How many solar cells would it take to provide the electricity for my house? • How much coal would it take to light a 100-watt light bulb 24 hours a day for 1 year? • How are torpedoes propelled through the water?

Is it possible to generate electricity directly from heat?

If you have a lot of heat, then you can do what power plants do — you can use the heat to generate steam, and use the steam to spin a turbine. The turbine can drive a generator, which produces electricity. This setup is very common, but it requires a fair amount of equipment and space.

If you want to generate electricity from heat in a simple way that involves no moving parts, the most routine way to do this involves *thermocouples*. Thermocouples take advantage of an electrical effect that occurs at junctions between different metals. For example, you could take two iron wires and one copper wire. Twist one end of the copper wire and one end of one of the iron wires together. Do the same with the other end of the copper wire and the other iron wire. If you heat one of the twisted junctions (for example with a match) and attach the two free ends to a voltmeter, you will be able to measure a voltage, which means that electricity is being generated through the wires.

Interplanetary satellites that fly toward planets like Jupiter and Saturn are so far away from the sun that they cannot use solar panels to generate electricity. These satellites use radioisotope thermoelectric generators (RTGs) to generate their power. An RTG uses radioactive material (like plutonium) to generate heat, and thermocouples convert the heat to electricity. RTGs have no moving parts, so they are reliable, and the radioactive material generates heat for many years so they last a long time.

⚙ **Web Links**

How Satellites Work

Is flour flammable? I heard if you were to burn flour, it would explode. If so, why does this happen?

TOP 20

White flour is made up mostly of starch. Starch is a carbohydrate, meaning that it is made of sugar molecules chained together. Anyone who has ever lit a marshmallow on fire knows that sugar burns easily. So does flour.

Flour and many other carbohydrates become explosive when they are hanging in the air as dust. It takes only 1 or 2 grams of dust per cubic foot of air (50 or so grams per cubic meter) for the mixture to be ignitable. Flour grains are so tiny that they burn instantly. When one grain burns, it lights other grains near it, and the flame front can flash through a dust cloud with explosive force. Just about any carbohydrate dust — including sugar, pudding mix, and fine sawdust — will explode once ignited.

When you hear about an explosion in a grain elevator, this is what has occurred. A spark or a source of heat has ignited the dust in the air and it exploded.

○ **Web Links**

How Food Works

How does an oxygen canister work?

When we think of storing oxygen, what we usually think about is large metal tanks holding pressurized oxygen gas. This is the way we see oxygen in hospitals and on welding rigs. We also see scuba divers taking their oxygen with them in the form of compressed air in scuba tanks. Because oxygen tanks are so common, we tend to think that heavy metal tanks are the only way to store oxygen.

In addition, however, there is a chemical way to store oxygen. Many chemicals, including potassium chlorate ($KClO_3$) and sodium chlorate ($NaClO_3$) are rich in oxygen and willingly give it up as a nearly pure gas when heated. The scuba tanks that divers wear might weigh up to 80 pounds (36 kilograms) but can store only a few hours of air. An oxygen canister weighing 33 pounds (15 kilograms) can provide 4 days worth of oxygen. The sodium chlorate is acting something like an oxygen sponge, and you squeeze the oxygen out with heat.

Modern oxygen canisters are a lightweight way to store oxygen. You find oxygen canisters (also known as chemical oxygen generators) on airplanes, submarines, and space stations — places where oxygen can run out unexpectedly. Typically an oxygen canister contains a sodium chlorate pellet or cylinder and an igniter. The igniter generates enough heat to start the sodium chlorate reaction, and then the heat of the reaction sustains itself. The sodium chlorate does not burn; its decomposition just gives off lots of heat and lots of oxygen.

Oxygen canisters can cause fires because they are hot and they generate oxygen. Anything nearby that happens to ignite will burn intensely because of the rich oxygen supply.

What home appliances use the most power?

If you were to make a chart of the electricity-consuming devices in a typical home and rank them in order of their hunger for power, the list might look something like this:

Device	Typical Consumption	Cost per hour
Heat pump or central air	15,000 watts	$1.50
Water heater or clothes dryer	4,000 watts	40 cents
Water pump (well)	3,000 watts	30 cents
Space heater	1,500 watts	15 cents
Hair dryer	1,200 watts	12 cents
Electric range burner	1,000 watts	10 cents
Refrigerator	1,000 watts	10 cents
Computer and monitor	400 watts	4 cents
Light bulb	60 watts	0.6 cents

This table assumes that a kilowatt-hour of electricity costs 10 cents.

If your house has electric heat, you will use a lot of power in the middle of winter. A heat pump might run 10-15 hours a day. At $1.50 an hour, that's $15 to $22 per day. Over the course of a month, you will use several hundred dollars worth of electricity. You will incur the same costs in the summer if you use the air conditioner a lot.

Water heating uses a good bit of power as well. When you take a shower or run a load of clothes in the washer, the electric water heater might run for an hour re-heating the water in the tank. That's 40 cents. A typical household can burn several dollars each day heating water.

Refrigeration is another big power drain because the refrigerator can easily run for 10 hours a day. That's about $1 per day. If you leave the computer or TV on all day it can add up to $1 per day as well.

Using a space heater or an electric blanket to heat a smaller area at night is probably the easiest way to save big on your power bill. Saving hot water is the next easiest.

○ Web Links

How Emergency Power Systems Work
How Power Distribution Grids Work

Do commercial jets have locks on the doors and ignition keys?

Small planes have locks on the doors and ignition keys inside to start the engine. They are pretty much like a car in this respect. The reason these planes need locks and keys is because they sit out in relatively unsecured parking areas. They need locks for the same reason that a car does when it is sitting in a parking lot.

Commercial jets, on the other hand, have no locks on the doors and no ignition key of any sort. You can hop in, flip a couple of switches and start one up!

Jumbo jets don't need an ignition key for the same reason that you probably do not lock your car when it is sitting in your

Power Up!

○

locked garage. Because your garage is secure, you have no need to secure your car. Security at an airport is generally very good, and planes are kept locked in hangars or attached to jetways that are secure and under constant surveillance. In addition, if you did somehow make your way into the jet and start it up, the chance of your being caught is 100%, which provides a good deterrent.

🔘 Web Links

How Airplanes Work

My lawn mower doesn't have a battery like a car does, so where does the electricity to spark the spark plug come from?

Most small lawn mowers, chain saws, trimmers, and other small gasoline engines do not need a battery. Instead, they generate the power for the spark plug by using a magneto. Magnetos are also used on many small airplanes because they are extremely reliable.

The idea behind any ignition system is to generate an extremely high voltage — on the order of 20,000 volts — at exactly the right time. The voltage causes a spark to jump across the spark plug's gap, and the spark ignites the fuel in the engine.

In a magneto, a strong magnet is moved past a coil of wire to generate electricity.

A magneto consists of five parts:

- An armature — a U-shaped piece of iron
- A primary coil of perhaps 200 turns of thick wire wrapped around one leg of the U
- A secondary coil of perhaps 20,000 turns of very thin wire wrapped around the primary coil
- A simple electronic control unit that is commonly called an electronic ignition (or a set of breaker points and a capacitor)
- A pair of strong permanent magnets embedded in the engine's flywheel

When the magnets fly past the U-shaped armature, they induce a magnetic field in the armature. This field induces some small amount of current in the primary and secondary coil. What you need to spark the spark plug, however, is extremely high voltage. Therefore, as the magnetic field in the armature reaches its maximum, a switch in the electronic control unit opens. This switch breaks the flow of current through the primary coil and causes a voltage spike of perhaps 200 volts. The secondary coil, having 100 times more turns than the primary coil, amplifies this voltage to approximately 20,000 volts, and this voltage feeds to the spark plug.

Many riding lawn mowers do have a battery if they have accessories like headlights and an electric starter. Even so, the engine may use a magneto because the magneto is so simple and reliable.

○ Web Links

How Car Engines Work
How Chain Saws Work
How Electromagnets Work
How Two-Stroke Engines Work

How does a laser speed gun work to measure a car's speed compared to normal police radar?

A traditional radar set sends out a radio pulse and waits for the reflection. Then it measures the Doppler shift in the signal and uses the shift to determine a vehicle's speed.

Laser speed guns use a more direct method that relies on the reflection time of light rather than Doppler shift. You have probably experienced the reflection time of sound waves in the form of an echo. For example, if you shout down a well or across a canyon, the sound takes a noticeable amount of time to reach the bottom of the well and travel back to your ear. Sound travels at something like 1,000 feet (300 meters) per second, so a deep well or a wide canyon creates a very apparent round-trip time for the sound.

A laser speed gun measures the round-trip time for light to reach a car and reflect back. Light from a laser speed gun moves a lot faster than sound — about 984 million feet (300 million meters) per second, or roughly 1 foot (30 centimeters) per nanosecond. A laser speed gun shoots a very short burst of infrared laser light and then waits for the light to reflect off the vehicle. The gun counts the number of nanoseconds it takes for the round trip, and by dividing by 2 it can calculate the distance to the car. If the gun takes 1,000 samples per second, it can compare the change in distance between samples and calculate the speed of the car.

The advantage of a laser speed gun (for the Police, anyway) is that the size of the "cone" of light that the gun emits is very small, even at a range like 1,000 feet (300 meters). The cone at this distance might be 3 feet (1 meter) in diameter, which allows the gun to target a specific vehicle. A laser speed gun is also very accurate. The disadvantage is that a police officer has to aim a laser speed gun; normal police radar with a broad radar beam can detect Doppler shift without an officer aiming the radar.

⚙ Web Links

How Lasers Work
How Radar Works

How many solar cells would it take to provide the electricity for my house? **TOP 20**

A typical solar cell that is 1 inch (6.45 centimeters) square generates 0.45 volts and 100-200 milliamps, or 45-90 milliwatts. For the sake of discussion, let's assume that a panel can generate 70 milliwatts per square inch.

To calculate how many square inches of solar panel you need for a house, you need to know:

- How much power the house consumes on average.
- Where the house is located — so that you can calculate factors such as mean solar days, average rainfall, and so on. This question is impossible to answer unless you have a specific location in mind. We'll assume that on an average day the solar panels generate their maximum power for 5 hours.

A "typical home" in America can use either electricity or gas to provide heat for things like heating the house, heating the hot water, or running appliances like the clothes dryer, stove, or oven. Even if you were to power a house with solar electricity, you would probably also want to use gas appliances because solar electricity is so expensive. This means that what you would be powering with solar electricity are devices like the refrigerator, the lights, the computer, the TV, stereo equipment, and motors in things like furnace fans and the washing machine. Let's say that all of this equipment averages out to 600 watts. Over the course of 24 hours, you need:

600 watts x 24 hours = 14,400 watt-hours per day

From our calculations and assumptions, we know that a solar panel can generate:

70 milliwatts per square inch x 5 hours = 350 milliwatt hours per day

Therefore you need about 41,000 square inches of solar panel for the house. That's a solar panel that measures 17' x 17' or so (5.18 meters square). Currently, this panel would cost around $16,000. Then, because the sun only shines part of the time, you would need to purchase a battery bank, an inverter, and other accessories, which often doubles the cost of the installation.

If you want to have a small room air conditioner in your bedroom, double everything.

Because solar electricity is so expensive, you would normally go to great lengths to reduce your electricity consumption. Instead of a desktop computer and a monitor you would use a laptop computer. You would use fluorescent lights instead of incandescent. You would use a small black-and-white TV instead of a large color set. You would get a small, extremely efficient refrigerator. By doing these things you might be able to reduce your average power consumption to 100 watts. This would cut the size of your solar panel and its cost by a factor of 6, which would make the expense more reasonable.

At this point, 100 watts per hour purchased from the power grid costs only about 24 cents per day, or $91 a year. That's why you

Power Up!

don't see many solar houses except in very remote locations. When it only costs about $100 a year to purchase power from the grid, it is hard to justify spending thousands of dollars on solar panels.

○ Web Links

How Solar Yard Lights Work

How much coal would it take to light a 100-watt light bulb 24 hours a day for 1 year?

In a year, a 100-watt light bulb uses 0.1 kW x 8,760 hours, or 876 kWh.

The thermal energy content of coal is 6,150 kWh/ton. Although coal-fired power generators are very efficient, they are still limited by the laws of thermodynamics. Only about 40% of the thermal energy in coal is converted to electricity. So the electricity generated per ton of coal is 0.4 x 6,150 kWh, or 2,460 kWh/ton.

So a light bulb uses 714 pounds (325 kilograms) of coal in a year. That is a pretty big pile of coal, but let's look at what else was produced to power that light bulb.

A typical 500-megawatt coal power plant produces 3.5 billion kWh per year. That is enough energy for 4 million light bulbs to operate year round. To produce this amount of electrical energy, the plant burns 1.43 million tons of coal.

Pollutant Total for One Light	Power Plant	Bulb-Year's Worth
Sulfur Dioxide — Main cause of acid rain	10,000 tons	5 pounds
Nitrogen Oxides — Causes smog and acid rain	10,200 tons	5.1 pounds
Carbon Dioxide — Greenhouse gas suspected of causing global warming	3,700,000 tons	1,852 pounds

It also produces smaller amounts of just about every element on the periodic table, including the radioactive ones. In fact, a coal-burning power plant emits more radiation than a (properly functioning) nuclear power plant.

○ Web Links

How Nuclear Power Works
How the Power Distribution Grid Works

How are torpedoes propelled through the water?

A torpedo is essentially a guided missile that happens to "fly" underwater. A torpedo therefore has a propulsion system, a guidance system, and some sort of explosive device. Torpedoes can travel several miles on their way to the target, so they need a propulsion system that can run for 10-20 minutes.

Most missiles that fly through the air use either rocket engines or jet engines, but neither of these work very well underwater. Torpedoes use one of two techniques for propulsion:

- Batteries and an electric motor. This is the same technique that any non-nuclear submarine must use when running underwater.

- Engines that use special fuel. Most engines that we are familiar with, like car engines and jet engines, draw their oxygen from the air around the engine and use it to burn a fuel. A torpedo cannot do that, so it uses a fuel that either does not need an oxidizer or that carries the oxidizer inside the torpedo. Otto fuel, which is a common fuel for U.S. torpedoes, has its own oxidizer mixed with the fuel.

Otto fuel is something like nitroglycerine. Nitroglycerine is explosive, and it is also very unstable. The main component in Otto fuel is propylene glycol dinitrate, which is chemically similar to nitroglycerine but much more stable. When ignited with a spark, Otto fuel burns but it needs no oxygen because the oxygen is a part of the chemical structure of propylene glycol dinitrate.

Power Up!

○

69

We don't often encounter fuels that contain their own oxidizers. When a fuel has its own oxidizer, it tends to be explosive. Dynamite, for example, has its own oxidizer.

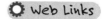

How Cruise Missiles Work

Stuff Around Your House

⚙ How do pop-up turkey timers work? • What is the bumpy stuff on my ceiling that looks like popcorn or cottage cheese? • How do electric stud finders know the location of the studs? • How does pressure-treated lumber work? What does "pressure-treated" mean? • Why do the two flat prongs on the plugs for electrical appliances have holes in them? • What is the difference between analog and digital cell phones? • How does caller ID work? • How do the lamps that you can touch to turn them on work? • How do radio signals from the National Atomic Clock in Colorado reach mine?

How do pop-up turkey timers work?

The pop-up timer is a simple piece of technology that lets everyone know when the turkey or chicken is done — when the red indicator stick pops up, it's time to eat!

A turkey, for example, is finished cooking when it reaches 185°F (85°C); at this point, the pop-up timer pops up to tell you that the bird is fully cooked. If you don't use a pop-up timer, you can use a meat thermometer to figure out the temperature of the meat. But the pop-up timer, which often comes already inserted into turkeys available at the grocery store, can be more fun to use. Unlike a thermometer, the timer is binary; you'll have no trouble reading it because the answer is either "Yes" or "No."

A pop-up timer found in a turkey or chicken normally has four parts:

- The outer case (typically white or light blue)
- The little stick that pops up (typically red)
- A spring
- A blob of soft metal similar to solder

The soft metal is solid at room temperature and turns to a liquid (melts) at about 185°F (85°C). When the metal turns to a liquid, it frees the end of the red stick that had been trapped in the metal. The spring pops the red stick up and you know the turkey is done!

One little-known fact is that these timers are reusable. If you dip the tip in boiling water it will re-melt the metal and you can push the pop-up piece back into the metal. Then let it cool and the pop-up piece will be back in its original position, ready to use again!

What is the bumpy stuff on my ceiling that looks like popcorn or cottage cheese?

The ceilings of many homes and apartments are not painted, but instead are coated with a substance that has a bumpy texture.

Builders normally call this treatment a textured finish or an acoustic finish, but most people refer to it as popcorn. This sort of ceiling treatment is popular for several reasons:

- It keeps you from having to plaster and sand the ceiling. If you've ever plastered a ceiling, you know that this task is no fun.
- It hides lots of imperfections in the ceiling. The ceiling is a huge, flat, uninterrupted, and well-lit expanse. Any imperfection is immediately obvious. The texturing hides imperfections very effectively.
- It helps eliminate echo in a room. If you have ever talked in a room before and after carpeting, you know how much installing carpeting will decrease echoes in a room. An acoustic finish is like carpeting the ceiling.

There are many different ways to put a texture on a ceiling, but the most common technique involves mixing some sort of lumpy aggregate — either vermiculite or polystyrene — with ceiling paint. The mixture is then sprayed onto the ceiling using a special spray gun. This paint goes on in one coat and once it dries — like normal paint — you're done.

How do electric stud finders know the location of the studs?

Whether you are hanging pictures, putting up a new set of shelves, or getting ready to start an addition, knowing where the studs are in a wall can be extremely handy. Before there were stud finders, either you pounded a small nail into the wall until you hit a stud, or you used a small pivoting magnet, which would help you find nails that had been driven into a stud.

Electronic stud finders changed all that. They give you an amazingly accurate view into the wall and show you exactly where each stud is. They use changes in capacitance to sense where the stud is. When the plate inside the stud finder is positioned over wallboard, it will sense one dielectric constant; when it is positioned over a stud, the dielectric constant is different. The circuit in the stud finder can sense the change and reports it on its display.

The latest technology in stud finders uses a very small radar system to detect the stud.

○ Web Links

How Radar Works

How does pressure-treated lumber work? What does "pressure-treated" mean?

Wood is a great building material. It is strong, lightweight, easily worked with tools, and relatively inexpensive. The only problem with wood is that many varieties of bacteria, fungi, and insects find it appetizing. When wood is in contact with the ground or moisture for any period of time, these organisms attack the wood. Untreated wood like pine will only last a year or two if it is touching moist ground.

Pressure-treated lumber is wood that has been immersed in a liquid preservative and placed in a pressure chamber. The chamber forces the chemical into the wood fibers. The pressurized approach makes sure the chemical makes it to the core of each piece of wood — it is much more effective than simply soaking the wood in the chemical.

The most common chemical used to treat lumber is called chromated copper arsenate, or CCA. Copper and arsenic are both toxic to different types of organisms that attack wood. The chromium helps to bond the copper to the wood to prevent leaching. CCA binds to wood fibers very well and allows wood to last decades when it is in contact with the ground.

The protection provided by the chemical depends on the amount of chemical that the wood absorbs. In the U.S., the amount of chemical is measured in pounds of chemical per cubic foot of wood. For ground contact, 0.40 pounds per cubic foot are needed. For foundations 0.60 pounds per cubic foot is the standard.

The chemicals in treated wood are generally not good for humans either. This is why you will see warnings advising you to wear gloves, avoid breathing the sawdust, and refrain from burning treated wood. Keeping small children away from treated wood is also appropriate.

Why do the two flat prongs on the plugs for electrical appliances have holes in them?

If you unplug any appliance in your house, there's a 98% chance that the two flat prongs have holes in them. The obvious question to ask is "why?"

There are two reasons for the holes:

- If you were to take apart an outlet and look at the contact wipers that the prongs slide into, you would find that they have bumps on them. These bumps fit into the holes so that the outlet can grip the plug's prongs more firmly. This detenting prevents the plug from slipping out of the socket due to the weight of the plug and cord. It also improves the contact between the plug and the outlet.

- Electrical devices can be "factory-sealed" or "locked-out" by the manufacturer or owner using a plastic tie or a small padlock through one or both of the plug prong holes. Construction projects or industrial safety requirements may require this type of sealing. For example, a manufacturer might apply a plastic band through the hole and attach it to a tag that reads, "You must do blah blah blah before plugging in this device." The user cannot plug in the device without removing the tag, so the user is sure to see the tag.

What is the difference between analog and digital cell phones?

The three most popular cellular services in the U.S. are normal analog cell phones (AMPS), digital cell phones, and personal communications services (PCS).

An analog cell phone transmits and receives analog radio signals in much the same way that a CB radio does. The voice signal isn't encoded or compressed in any way. This is why it's possible for people with radio scanners to pick up analog cell phone conversations. This service is known as AMPS.

There are two popular digital systems. One is generically known as digital service, and the other is clearly branded as PCS. Digital service uses newer, digital phones, but they communicate with normal AMPS towers. A call is set up using the normal AMPS protocol, and then the conversation is transmitted digitally.

Digital phones chop a single AMPS voice channel into three digital channels using a technique called Time Division Multiple Access (TDMA). That means that three phones can share the same channel, and they each understand that they will only use the channel part of the time. One phone uses the channel for a subset of a second, then the second phone uses it, then the third, and the cycle repeats. Digital cell phones, as you can see, are a hybrid between the existing analog system and digital technology.

PCS phones are completely digital and they use a whole separate set of towers at a whole separate set of frequencies between 1.85 and 2.15 GHz. Because of the higher frequencies, there must be more towers closer together. Because the service is completely digital, encryption and error-correcting codes can make the call much clearer and nearly impossible to intercept. PCS normally bundles in other services like paging, caller ID, and even e-mail.

How a Cell Phone Works

How does caller ID work?

If you have a caller ID box attached to your phone, an amazing thing happens every time your phone rings: The number (and sometimes even the name) of the calling party appears on the display immediately following the first ring.

The process of making the caller ID display possible is remarkably simple at your end of the line. Early modems used a technique called Frequency Shift Keying (FSK) to transmit bits over a phone line. One tone (or frequency, like 1,200 Hz) represents a binary 1, while another tone (like 2,200 Hz) represents a binary 0. A modem changes frequencies depending on whether it wants to send a 1 or a 0. How quickly it changes frequencies determines the speed, or baud rate, of the modem.

To send caller ID information to your home, the phone company uses an FSK technique identical to a 1,200 baud modem, and it sends ASCII character data to the caller ID box. The modem message is sent between the first and second ring. So the phone rings once, and if you could listen to the phone line just after that ring, you would hear a "bleeeep" sound about half a second long. If you decoded the bleep, you would find that it contains:

- A series of alternating 1s and 0s to help the caller ID box get the timing down
- A series of 180 1s
- A byte representing the type of message
- A byte representing the length of the message
- Month, day, hour, and minute, each represented with a pair of bytes
- The 10-digit phone number in 10 bytes
- A checksum byte

A more advanced system contains other information such as the caller's name, but its technique is identical. Each character is sent as a standard 8-bit ASCII character preceded by a "0" start bit and followed by a "1" stop bit.

The caller ID box contains a modem to decode the bits, a little circuit to detect a ring signal, and a simple processor to drive the display.

Stuff Around Your House

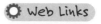

How Telephones Work

How do the lamps that you can touch to turn them on work?

Switches that are sensitive to human touch — as opposed to switches that must be flipped or pushed to make and break a mechanical connection — have been around for many years. The most important advantage of these lamps is that dirt and moisture cannot get into the switch to gum it up or damage it. Over the years, many different properties of the human body have been used to flip touch-sensitive switches:

- **Temperature.** The human body is generally warmer than the surrounding air. Many elevators therefore use buttons that are sensitive to the warmth of the human finger. These buttons, of course, don't work if you have cold hands. They also have the unfortunate property of taking an elevator to a floor where a fire is burning.

- **Resistance.** The human body, being made mostly of water, conducts electricity fairly well. By placing two contacts very close together, your finger can close the circuit when you touch it.

- **Radio reception.** You may have noticed that when you touch an antenna, the reception gets better on a TV or radio. That's because the human body makes a pretty good antenna. There are even small liquid crystal display (LCD) TVs that have a conductive neck strap so that the user acts as the antenna! Some touch-sensitive switch designs simply look for a change in radio-wave reception that occurs when the switch is touched.

Touch-sensitive lamps almost always use a fourth property of the human body: its *capacitance*. The word capacitance has as its root the word capacity. Capacitance is the capacity an object has to hold electrons. The lamp, when standing by itself on a table, has a certain capacitance. This means that if a circuit tried to charge the lamp with electrons, a certain number would be needed to "fill" it. When you touch the lamp, your body adds to its capacity. It takes more electrons to fill you and the lamp, and the circuit detects that difference.

How do radio signals from the National Atomic Clock in Colorado reach mine?

Many gadget catalogs and high-tech stores sell "radio-controlled" clocks and wristwatches that are able to receive these radio signals. These clocks and watches truly are synchronizing themselves with the atomic clock in Colorado. This feature is made possible by a radio system set up and operated by the National Institute of Standards and Technology (NIST), located in Fort Collins, Colorado. NIST operates radio station WWVB, which is the station that transmits the time codes.

WWVB is a very interesting radio station. It has high transmitter power (50,000 watts), a very efficient antenna, and an extremely low frequency (60,000 Hz). For comparison, a typical AM radio station broadcasts at a frequency of 1,000,000 Hz. The combination of high power and low frequency gives the radio waves from WWVB a lot of bounce, and this single station can therefore cover the entire continental United States plus much of Canada and Central America as well.

The time codes are sent from WWVB using one of the simplest systems possible, and at a very low data rate of one bit per second. For comparison, a typical modem transmits over the phone lines at tens of thousands of bits per second; imagine receiving a Web page at one bit per second! The 60,000 Hz signal is always transmitted, but every second it is significantly reduced in power for a period of 0.2, 0.5 or 0.8 seconds:

- 0.2 seconds of reduced power means a binary 0
- 0.5 seconds of reduced power is a binary 1
- 0.8 seconds of reduced power is a separator

The time code is sent in BCD (Binary Coded Decimal) and indicates minutes, hours, day of the year, and year, along with information about daylight savings time and leap years. The time is transmitted using 53 bits and 7 separators, and therefore takes 60 seconds to transmit.

A clock or watch can contain an extremely small and relatively simple antenna and receiver to decode the information in the signal and set the clock's time accurately. To take advantage of this clock, all you have to do is set the time zone; the clock can then display a very accurate time. The only thing more accurate that you can carry around easily is a global positioning satellite (GPS) receiver, which derives atomic clock accuracy in real time from the atomic clocks in orbiting GPS satellites.

?

8

Foodstuffs

⚙ How does popcorn work? • How much sugar is in soft drinks? • Why is root beer called root beer? • How is cotton candy made? • What is carrageenan? • Why shouldn't dogs eat chocolate? • How does that plastic ball-shaped widget inside beer release gas to aerate the beer? • How is the caffeine removed from coffee beans? • What is mayonnaise and how is it made?

How does popcorn work?

Popcorn certainly is unique. You toss a flat pouch no larger than a wallet into a microwave oven and in 3 minutes it has expanded to a volume 40 or 50 times bigger. Not too many other foods act this way!

Three elements make popcorn work as it does:

- Moisture inside the kernel
- Starch inside the kernel
- The hard shell surrounding the kernel

When a popcorn kernel heats up (either in a popcorn popper or in the microwave), the moisture inside the kernel expands. Moisture, it turns out, is extremely important to a popcorn kernel. Unless the percentage of moisture in the kernel is just right, the kernel won't pop. When the pressure inside the kernel gets high enough, the kernel explodes. That part seems normal enough — lots of things explode when you heat them.

The strange part is the white solid that forms during the process. The fluffy white stuff you like to eat is actually the starch that was inside the kernel. As the moisture expands inside the kernel, so does the starch. It forms a cluster of bubbles much like soapsuds in a bubble bath. Two everyday items you're familiar with behave similarly. The first is bread or muffins, which expand and solidify (although much more slowly). The other is "silly string," which is a liquid that solidifies as soon as it hits air.

Here are three experiments you can perform to get a better understanding of how popcorn works:

- Use a needle or pushpin to puncture the shells of several popcorn kernels. Be sure to wear glasses when you do this — it is not as easy as it sounds! Then try to pop the kernels. They won't pop because the pressure cannot build inside the punctured kernels.
- Let the kernels stand in a warm oven or in the sun for several days, and then try popping them. The oven or sun will dry out the kernels and make them difficult to pop.
- Try to pop popcorn at a low temperature (below 300°F or 150°C). You will find that the popcorn will not pop; it has to get beyond a certain temperature for the pressure inside the kernel to build to the point where popping occurs.

How much sugar is in soft drinks?

There's probably more sugar in a typical can of soda than you think! Here are two ways to prove it to yourself.

The first is to buy a packet of Kool-aid or Flavor-aid unsweetened soft drink mix — the kind you add sugar to when you are making it. You will be instructed to add one cup of sugar and enough water to make two quarts (64 ounces). A cup of sugar contains 48 teaspoons of sugar. Therefore, a 16-ounce (473 milliliters) serving of one of these beverages contains 12 teaspoons (59 milliliters) of sugar!

The other way to prove it is to look at the calorie count of any soft drink. A typical carbonated soft drink has 200 calories in a 16-ounce serving. All of those calories come from sugar, and sugar contains 16 calories per teaspoon. By this measurement, a 16-ounce serving (473 milliliters) contains 12.5 teaspoons (61.6 milliliters) of sugar.

Go to the kitchen and get out a 16-ounce glass, a teaspoon, and some sugar. Measure 12 teaspoons of sugar into the glass — it's an astonishing amount! Then multiply that by the number of sodas you typically consume in a day.

Why is root beer called root beer?

The English language has lots of words that are used in two or three different ways. For example, the word *cabinet* can mean "storage space in your kitchen" or "a group of folks who advise the president." Beer is a word with two meanings. It can mean an alcoholic beverage made from cereal grains, or a non-alcoholic beverage flavored by root extracts. Root beer, birch beer, and ginger beer are three common forms of this second type of beer.

In the case of root beer, the flavoring comes from the root of the sassafras tree or the sarsaparilla vine. Originally, the root was brewed like a tea to make an extract, but now it is much easier to buy the extract ready made.

The root beer extract is mixed with sugar, yeast, and water. If this mixture is placed in a tightly sealed bottle, the yeast will generate carbon dioxide at a high enough pressure to carbonate the water. You end up with fizzy, delicious root beer!

Beer and root beer are made of the same active ingredients: sugar, yeast, and water. So, why does beer contain alcohol, but root beer does not?

When you make beer, you allow the yeast to ferment at normal atmospheric pressure for several weeks, until most of the sugar is consumed by the yeast. That is why beer is bitter rather than sweet. With root beer, you bottle it right after you mix it up. The yeast is able to develop pressure in the bottle and carbonate the root beer, but then the reaction stops. Root beer is therefore sweet and contains virtually no alcohol.

O Web Links

How Beer Works

How is cotton candy made?

Cotton candy is one of those intriguing foods that makes no sense until you know the secret. You cannot produce cotton candy without special equipment, but if you have the equipment, it is incredibly easy to make.

Cotton candy is nothing but pure sugar. To make the sugar "cottony," you need four things:

- Heat to melt the sugar and turn it into a liquid
- A set of very small holes that the liquid sugar can flow through to form threads of sugar
- A spinning head that slings the liquid sugar outward so it is forced through the holes
- A bowl to catch the threads

First, sugar is poured into the center of the head. The head contains the heater to melt the sugar and make it liquid. Then, by spinning the head, the cotton candy machine forces the liquid sugar out through tiny holes in the head. The instant the thin threads of sugar hit the air, they cool and re-solidify, so a web of sugar threads develops in the bowl of the machine. The web is easily collected on a paper cone.

If you mix water with cotton candy, it instantly dissolves, and you can see that there is not more than a tablespoon or two of sugar filling the entire bag. It's amazing how much a vendor can charge for sugar, air, and a little coloring!

What is carrageenan?

Lots of foods can contain some pretty weird-sounding stuff. That's because processed foods must do some remarkable things. Imagine that a cookie is made in Texas, is trucked across the country in the middle of the summer, sits in a warehouse for a couple of weeks, and then rides home in the trunk of your car. And when you open the package, you want the cookie to look perfect! Not an easy thing, it turns out...

Food items such as liquids and cheese can be even more problematic than our example cookie. Their natural inclination is to separate, foam, melt, and precipitate, especially after bouncing down the road for a thousand miles. That's why many foods contain chemicals known as gums. Two gums everyone is familiar with are gelatin and corn starch. If you look at processed food, you see all sorts of other gums like carrageenan, xanthan gum, cellulose gum, locust bean gum, agar, and so on. Food scientists (not cooks — food scientists make processed foods) use these substances for four main reasons:

- **They thicken things.** Ice cream, marshmallow fluff, pancake syrup, and similar items all benefit from thickening.
- **They emulsify things.** They help liquids to stay mixed together without separating.
- **They change the texture.** Generally a gum will make something thicker or chewier and give it more texture.
- **They stabilize crystals.** A gum might help prevent sugar or ice from crystallizing.

These are all handy capabilities when making food products that have to look good for several months after traveling across the country. The reason a normal cook usually does not need to use things like carrageenan or xanthan gum is because the food a cook makes gets eaten quickly and is not mistreated. A cook can also use less expensive ingredients such as gelatin, flour, or eggs because the time span between cooking and consumption is so short.

Carrageenan, by the way, is a seaweed extract. This particular type of seaweed is common in the Atlantic Ocean near Britain, Europe, and North America. You boil the seaweed to extract the carrageenan. In that sense carrageenan is completely "natural" — it's not much different from tomato paste in its creation.

Why shouldn't dogs eat chocolate?

A chemical in chocolate called theobromine is the source of the problem. Theobromine is similar to caffeine and is toxic to a dog when the dog ingests between 100-150 milligrams per kilogram of body weight, or 45-70 milligrams per pound. Different types of chocolate contain different amounts of theobromine, so it would take 20 ounces of milk chocolate to kill a 20-pound dog, but only 2 ounces of baker's chocolate or 6 ounces of semisweet chocolate. It's not hard for a dog to get into an Easter basket full of chocolate eggs and bunnies and gobble up a pound or two of chocolate. If the dog is small, that binge could be deadly.

Actually, chocolate poisoning is not as unusual as it sounds. For a human being, caffeine is toxic at levels of 150 milligrams per kilogram, or 70 milligrams per pound. That's exactly the same as dogs. Humans generally weigh a lot more than dogs, but small children can get into trouble with caffeine or chocolate if they consume too much too quickly. Infants are especially vulnerable because they don't eliminate caffeine from the bloodstream nearly as quickly as adults.

○ Web Links

How Chocolate Works

How does that plastic ball-shaped widget inside beer release gas to aerate the beer?

To answer this question, first you need to know a little bit about what makes beer fizzy and how a head forms.

Most beers are carbonated with carbon dioxide (CO_2). When the beer is in the can, some of this CO_2 is dissolved in the beer and some is at the top of the can. The CO_2 that is dissolved in the beer is what makes it fizzy. When the can is closed, the pressure inside the can is higher than the pressure outside, so that when you open the can, the sudden drop in pressure and the agitation of pouring causes some of the CO_2 to bubble out of the solution, forming a head on your beer.

A stout like Guinness has a creamier, longer-lasting head than a canned lager beer. In addition, Guinness is less fizzy than a regular lager beer. Guinness is canned with a mixture of carbon dioxide and nitrogen. Nitrogen is not absorbed into the beer nearly as well as carbon dioxide. Even though a can of Guinness may be at the same pressure as a can of lager, it contains less CO_2 (and is therefore less fizzy) because the nitrogen makes up some of the pressure.

Because a beer like Guinness contains less dissolved CO_2, if you poured it from a can with no widget, the head would not be very thick because most of the CO_2 would stay dissolved.

The purpose of the widget is to release the CO_2 from some of the beer in the can and create the head. The widget is a plastic, nitrogen-filled sphere with a tiny hole in it. The sphere is added to the can before the can is sealed. It floats in the beer, with the hole just slightly below the surface of the beer.

Just before the can is sealed, a small shot of liquid nitrogen is added to the beer. This liquid nitrogen evaporates during the rest of the canning process and pressurizes the can. As the pressure increases in the can, beer is slowly forced into the sphere through the hole, compressing the nitrogen inside the sphere.

When you open the can, the pressure inside immediately drops and the compressed gas inside the sphere quickly forces the beer out through the tiny hole into the can. As the beer rushes through the tiny hole, this agitation causes the CO_2 that is dissolved in the beer to form tiny bubbles that rise to the surface of the beer. These bubbles help form the head.

🔘 **Web Links**

How Beer Works

How is the caffeine removed from coffee beans?

Caffeine occurs naturally in more than 60 plants, including:

- The arabica plant, which produces coffee beans
- The Theobroma Cacao tree, which produces the beans that are the primary ingredient in chocolate
- Kola nuts, which are used in many cola drink products
- The Thea Sinensis plant, whose leaves are used for teas

When separated from its sources, caffeine is a white, bitter-tasting powder. It is useful as a cardiac stimulant and mild diuretic.

Several methods are used to remove caffeine from its natural sources:

- Methylene chloride processing
- Ethyl acetate processing
- Carbon dioxide processing
- Water processing

Methylene chloride is a chemical used as a solvent to extract caffeine from many raw materials. Molecules of caffeine bond to molecules of methylene chloride. The materials are softened in a water bath or in steam. Then the materials are processed with methylene chloride by either one of two methods:

- Using the "direct" method, caffeine is removed by directly soaking the materials in methylene chloride.
- Using the "indirect" method, caffeine, which is water soluble, is extracted by soaking the materials in water. Many of the flavors and oils are also extracted during this process, so the solution is treated with methylene chloride and then returned to the material for reabsorption of the flavorings.

Ethyl acetate processed products are referred to as "naturally decaffeinated" because ethyl acetate is a chemical found naturally in many fruits. Caffeine is extracted in the same way as with methylene chloride processing, but ethyl acetate is the solvent.

To decaffeinate using carbon dioxide (CO_2), water-softened materials are "pressure cooked" with the gas. At high pressures and high temperatures, carbon dioxide is in a supercritical state, acting as both a gas and a liquid. It becomes a solvent with its small, nonpolar molecules attracting the small caffeine molecules. Because flavor molecules are larger, they remain intact, which is why this process better retains the flavor of the material.

Caffeine extraction with water is used primarily for coffee decaffeination. The process is similar to the "indirect" method used in methylene chloride processing, but no chemicals are used. After the caffeine is leached out of the material by soaking in hot water for a period of time, the solution is passed through a carbon filter for caffeine removal. The water is then returned to the beans for reabsorption of flavors and oils. In the "Swiss Water Process," the same method is used, but instead of soaking in water, the beans are soaked in a coffee-flavored solution. This results in the caffeine being extracted without removing the coffee flavors.

Caffeine is not removed completely using any of these methods, but under federal regulations in the U.S., caffeine levels must not be above 2.5% of the product in order for a product to be labeled "decaffeinated."

Most of the caffeine removed in processing is manufactured for use in other products, such as medicines and soft drinks. For example, less than 5% of the caffeine found in cola drinks is actually from the kola nut, and many of the popular "high caffeine" soft drinks do not contain kola nut extracts at all. The caffeine content of soft drinks is primarily, and sometimes completely, from the addition of caffeine extracted from decaffeination processes.

○ Web Links

How Caffeine Works
How Chocolate Works

What is mayonnaise and how is it made?

Mayonnaise is a thick, creamy sauce or dressing that is made of oil, egg yolks, lemon juice or vinegar, and seasonings. Its not the same as salad dressing, which doesn't contain egg yolks and is sweeter than mayonnaise.

Mayonnaise is an *emulsion*, which is a mixture of one liquid with another that it normally can't combine with — oil and water is the classic example. Emulsifying is done by slowly adding one ingredient to another while simultaneously mixing rapidly. This action disperses and suspends tiny droplets of one liquid through another.

Even after mixing, the two liquids will quickly separate again if an emulsifier is not added. Emulsifiers are liaisons between the two liquids, and they stabilize the mixture. Among the foods that contain emulsifiers are eggs and gelatin. In mayonnaise, the emulsifier is egg yolk, which contains lecithin — a fat emulsifier.

Chemically, emulsions are colloids, which are heterogeneous mixtures composed of tiny particles suspended in another *immiscible* (unmixable) material. The particles are larger than molecules, but less than 1/1,000 of a millimeter (.001mm). Small particles like this do not settle out and pass right through filter paper. The particles in a colloid can be solid, liquid, or bubbles of gas. The medium in which they're suspended can be a solid, liquid, or gas (although gas-gas colloids aren't possible).

Emulsions are liquid-liquid colloids: tiny liquid droplets suspended in another liquid. Emulsions are usually thick in texture and satiny in appearance.

Emulsions are used in many different ways:

- By pharmacists as a vehicle for medication.
- In photography as a suspension of a salt of silver, platinum, and so on in gelatin or collodion used to coat plates, film, and paper.
- In explosives, paints, coatings, cosmetics, and detergents.
- In food, including baked goods and confectionary products.

Mayonnaise is made by combining lemon juice or vinegar with egg yolks. Eggs (containing the emulsifier lecithin) bind the ingredients together and prevent separation. Then, oil is added drop by drop as the mixture is rapidly whisked. Adding oil too quickly or insufficient rapid whisking will keep the two liquids from combining or emulsifying. As the sauce begins to thicken, however, oil can be added more rapidly. Seasonings are whisked in after all the oil has been added. Blenders, mixers, and food processors all enable you to easily make homemade mayonnaise, which many gourmands feel is far superior in taste and consistency to commercial mayonnaise.

Because homemade mayonnaise is uncooked, be sure to use the freshest eggs possible and ones that you are reasonably sure are free from salmonella. Homemade mayonnaise will last 3-4 days in the refrigerator.

Commercial mayonnaise, which will last up to 6 months in the refrigerator, usually contains modified food starch, cellulose gel, and other thickeners and emulsifiers. They also contain (by U.S. law) at least 65% oil by weight (except reduced-fat and fat-free mayonnaises).

Mayonnaise is used as the base for other sauces, such as tartar sauce and Thousand Island salad dressing. Aioli is garlic-flavored mayonnaise. Another classic emulsion sauce is hollandaise, which is a cooked mixture of butter, egg yolks, and lemon juice.

To Your Health

⚙ Why does the hair on my arms stay short, while the hair on my head can grow very long? • Why does hydrogen peroxide foam when I put it on a cut? • What does it mean when someone has 20/20 vision? • I was working in my garden and got a horrible case of poison ivy. What exactly causes this reaction? • How many senses does a person have? I always hear about five: touch, taste, smell, vision, and hearing. Do we have more? • What makes knuckles pop? • What constitutes a person's IQ? Does it improve with maturity, education, and experience? • How many calories does a person need daily? • What causes flatulence? • How does your stomach keep from digesting itself?

Why does the hair on my arms stay short, while the hair on my head can grow very long?

Each hair on your body grows from its own individual hair follicle. Inside the follicle, new hair cells form at the root of the hair shaft. As the cells form, they push older cells out of the follicle. As they are pushed out, the cells die and become the hair we see.

Depending on where the follicle is located on your body, it will produce new cells for a certain period of time. This period is called the growth phase. After the growth phase, the follicle will stop producing new cells for a period of time (the rest phase), and then restart the growth phase again. When the hair follicle enters the rest phase, the hair shaft breaks. So the existing hair falls out and a new hair takes its place. Therefore, the length of time that the hair grows during the growth phase controls the maximum length of the hair.

The cells that make the hairs on your arms are programmed to stop growing every couple of months, so the hair on your arms stays short. The hair follicles on your head, on the other hand, are programmed to let hair grow for years at a time, so the hair can grow very long.

Animals that shed have hair follicles that synchronize their rest phase. So all of the follicles enter the rest phase at once. Therefore, all the hair falls off at one time. A dog that sheds will lose its hair in large clumps. Many animals can also switch the coloring agent in the hair follicle on and off. So in the summer the hair is pigmented brown with melanin, but in the winter it is not, leaving the hair white in the winter.

Why does hydrogen peroxide foam when I put it on a cut?

You can buy hydrogen peroxide (H_2O_2) at the drugstore. What you are buying is a 3% solution, meaning the bottle contains 97% water and 3% hydrogen peroxide. Although most people use it as an antiseptic, it turns out that it is not very good as an antiseptic.

It is, however, not bad for washing cuts and scrapes, and the foaming looks cool.

Hydrogen peroxide foams because blood and cells contain an enzyme called catalase. Because a cut or scrape contains both blood and damaged cells, lots of catalase is floating around. When the catalase comes in contact with hydrogen peroxide, it turns the H_2O_2 into water (H_2O) and oxygen gas (O_2). Catalase does this extremely efficiently — up to 200,000 reactions per second per molecule. The bubbles you see in the foam are pure oxygen bubbles being created by the catalase. Try putting a little hydrogen peroxide on a cut potato and it will do the same thing for the same reason: Catalase in the damaged potato cells reacts with the hydrogen peroxide.

Hydrogen peroxide does not foam in the bottle or on your skin because there is no catalase to help the reaction to occur. Hydrogen peroxide is stable at room temperature.

○ **web Links**

How Cells Work

What does it mean when someone has 20/20 vision? ▚

By looking at lots of people, eye doctors have decided what a "normal" human being should be able to see when standing 20 feet away from an eye chart. If you have 20/20 vision, it means that when you stand 20 feet away from the chart, you can see what a normal human being can see. (In metric, the standard is 6 meters and it's called 6/6 vision.) In other words, if you have 20/20 vision, your vision is "normal," meaning that a majority of people in the population can see what you see at 20 feet.

If you have 20/40 vision, it means that when you stand 20 feet away from the chart, you can only see what a normal human can see when standing 40 feet from the chart. That is, if there is a "normal" person standing 40 feet away from the chart and you are standing only 20 feet away, you and the normal person can see the same detail. The term 20/100 means that when you stand 20 feet from the chart, you can only see what a normal person

○

standing 100 feet away can see. 20/200 is the cutoff for legal blindness in the United States.

You can also have vision that is better than the norm. A person with 20/10 vision can see at 20 feet what a normal person can see when standing 10 feet away from the chart.

Hawks, owls, and other birds of prey have much more acute vision than humans. A hawk has a much smaller eye than a human being but has lots of sensors (cones) packed into that space. This gives a hawk vision that is eight times more acute than a human's. A hawk might have 20/2 vision!

I was working in my garden and got a horrible case of poison ivy. What exactly causes this reaction?

Contact with a poison ivy plant can cause an uncomfortable and excruciatingly itchy rash. The rash is caused by a chemical called urushiol that is in the sap of the poison ivy plant. This chemical penetrates through the outer layer of skin until it hits the dermis, and in the dermis an allergic reaction to the urushiol occurs.

Given that description, there are a number of things you can deduce about poison ivy:

- **Not all people get poison ivy.** If your body does not mount an allergic reaction, then you can swim in urushiol and it will have no effect. It turns out, however, that the majority of people's immune systems react to urushiol after several exposures.

- **You cannot get poison ivy unless you come in contact with the sap that contains urushiol.** Keep in mind, however, that it is incredibly easy to come in contact with urushiol. You can get it from the plants directly. You can get it from taking off your shoes or pants if they have rubbed against poison ivy plants. You can get it from your dog or cat's fur if it walks through poison ivy.

- **The urushiol has to penetrate the skin to get to the dermis, so thin skin will show symptoms before thick skin will.**

- **Urushiol does not spread through the body (although it may appear to because of the delay in symptoms caused by differing skin thickness).** The blisters that form are also not contagious. They do not contain urushiol.

- **If you come in contact with poison ivy, washing off the sap will limit your reaction to it.** You have to wash it off before any significant penetration occurs.

The blisters that you see on your skin are the immune system's allergic reaction to urushiol. It takes 2-3 weeks for the skin to heal.

✪ Web Links

How Your Immune System Works

How many senses does a person have? I always hear about five: touch, taste, smell, vision, and hearing. Do we have more?

The standard list of five senses doesn't really give our bodies credit for all the amazing things they are able to do. We can actually sense at least a dozen different things.

In order for us to have a sense, we need a sensor. Each sensor is tuned to one specific sensation. For example, sensors in your eyes can detect light; that is all that they can detect. The easiest way to track down all the different senses a person has is to catalog the different sensors. Here is a reasonable list:

- In your eyes you have two different types of light sensors. One set of sensors, called the rods, sense light intensity and work well in low-light situations. The other type, called cones, can sense colors and require fairly intense light to be activated.

- In your inner ears there are sound sensors.

- Also in your ears are sensors that let you detect your orientation in the gravitational field; they give you your sense of balance.

- In your skin there are at least five different types of nerve endings:
  - heat sensitive
  - cold sensitive
  - pain sensitive
  - itch sensitive
  - pressure sensitive

These cells give us the sense of touch, sense of pain, sense of temperature, and sense of itch.

- In the nose there are chemical sensors that give us our sense of smell.
- On the tongue there are chemical receptors that give us our sense of taste.
- In the muscles and joints there are sensors that tell us where the different parts of our body are and the motion and tension of the muscles. These senses let us, for example, touch our index fingers together with our eyes shut.
- In the bladder there are sensors that indicate when it is time to urinate. Similarly the large intestine contains sensors that indicate when it is full.
- There is also the sense of hunger and thirst.

Depending on how you want to count it, there are between 14-20 different senses listed here.

Some people seem to have other senses. For example, many people can sense impending weather changes. Some mothers can sense when children are about to make a mess (also known as "eyes in the back of her head"). There is no scientific proof for any of these senses yet...

What makes knuckles pop?

If you've ever laced your fingers together, turned your palms away from you and bent your fingers back, you know what knuckle-popping sounds like. Joints produce that *CRACK* when bubbles burst in the fluid surrounding the joint.

Joints are the meeting points of two separate bones, held together and in place by connective tissues and ligaments. All of the joints in our bodies are surrounded by synovial fluid, a thick, clear liquid. When you stretch or bend your finger to pop the knuckle, you are causing the bones of the joint to pull apart. As they do, the connective tissue capsule that surrounds the joint is stretched. By stretching this capsule, you increase its volume. An increase in volume creates a decrease in pressure. So as the pressure of the synovial fluid drops, gases dissolved in the fluid become less soluble, forming bubbles through a process called *cavitation*. When the joint is stretched far enough, the pressure in the capsule drops so low that these bubbles burst, producing the pop that we associate with knuckle cracking.

It takes about 25-30 minutes for the gas to re-dissolve into the joint fluid. During this period of time, your knuckles will not crack. Once the gas is re-dissolved, cavitation is once again possible, and you can pop your knuckles again.

What constitutes a person's IQ? Does it improve with maturity, education, and experience?

IQ stands for intelligence quotient, a term derived from a scoring method developed by the German psychologist William Stern and first used in the Stanford-Binet Intelligence Scale. This and other early intelligence tests assigned subjects a "mental age" based on their test performance relative to the rest of a population. A subject who scored at the same level as an average 12-year-old, for example, was said to have a mental age of 12. The subject's overall intelligence score equaled the quotient of his mental age and his actual, chronological age, multiplied by 100:

(mental age/chronological age) x 100

So, a 10-year-old who scored the same as an average 12-year-old would have an intelligence quotient of 120:

(12/10) x 100 = 120

An intelligence quotient of 100 was considered average, since it was the score of someone whose mental age and chronological age were equal.

This method is no longer in general use because it has some significant flaws, chiefly that after childhood, raw scores don't proportionately increase with chronological age. Nowadays the term IQ generally describes a score on a test that rates the subject's cognitive ability as compared to the general population, using a standardized scale with 100 as the median score. On most tests, a score between 90 and 110, or the median plus or minus 10 indicates average intelligence. A score above 130 indicates exceptional intelligence and a score below 70 may indicate mental retardation. Like their predecessors, modern tests do take into account the age of a child when determining an IQ score. Children are graded relative to the population at their developmental level.

What is this cognitive ability being measured? Simply put, IQ tests are designed to measure your general ability to solve problems and understand concepts. This includes reasoning ability, problem-solving ability, ability to perceive relationships between things, and ability to store and retrieve information. IQ tests measure this general intellectual ability in a number of different ways. They may test:

- **Spatial ability.** The ability to visualize manipulation of shapes.
- **Mathematical ability.** The ability to solve problems and use logic.
- **Language ability.** This could include the ability to complete sentences or recognize words when letters have been rearranged or removed.
- **Memory ability.** The ability to recall things presented either visually or aurally.

Questions in each of these categories test for a specific cognitive ability, but many psychologists hold that they also indicate general intellectual ability. Most people perform better on one type of question than on others, but experts have determined that for the most part, people who excel in one category do similarly well in the other categories, and if someone does poorly in any one category, he also does poorly in the others. Based on this, these experts theorize there is one general element of intellectual ability that determines other specific cognitive abilities. Ideally, an IQ test measures this general factor of intelligence, abbreviated as *g*.

The best tests, therefore, feature questions from many categories of intellectual ability so that the test isn't weighted toward one specific skill.

Because IQ tests measure your ability to understand ideas and not the quantity of your knowledge, learning new information does not automatically increase your IQ. Learning may exercise your mind, however, which could help you to develop greater cognitive skills, but scientists do not fully understand this relationship. The connection between learning and mental ability is still largely unknown, as are the workings of the brain and the nature of intellectual ability. Intellectual ability does seem to depend more on genetic factors than on environmental factors, but most experts agree that environment plays some significant role in its development.

But can you increase your IQ score? Some evidence suggests that children develop higher intellectual ability if they receive better nurturing and diet as babies. Also, a higher degree of intellectual stimulation in preschool tends to boost children's IQ scores for a few years of elementary school but does not permanently increase IQ scores. For the most part, adult IQ scores don't significantly increase over time. There is evidence that maintaining an intellectually stimulating atmosphere (by learning new skills or solving puzzles, for example) boosts some cognitive ability, similar to the way maintaining an exercise regimen boosts physical ability, but these changes aren't permanent and do not have much effect on IQ scores.

So your IQ score is relatively stable, no matter what education you acquire. This does not mean that you can't increase your intelligence. IQ tests are only one imperfect method of measuring certain aspects of intellectual ability. A lot of critics point out that IQ tests don't measure creativity, social skills, wisdom, acquired abilities, or a host of other things we consider to be aspects of intelligence. The value of IQ tests is that they measure general cognitive ability, which has been proven to be a fairly accurate indicator of intellectual potential. There is a high positive correlation between IQ and success in school and the work place, but there are many, many cases where IQ and success do not coincide.

Web Links

How SATs Work

How many calories does a person need daily?

The number of calories that your body consumes in a day is different for every person. You may notice on the nutritional labels of the foods you buy that the "percent daily values" are based on a 2,000-calorie diet; 2,000 calories is a rough average of what people eat in a day. But your body might need more or less than 2,000. Height, weight, gender, age, and activity level all affect your caloric needs. Three main factors are involved in calculating how many calories your body needs per day:

- Basal metabolic rate
- Physical activity
- Thermic effect of food

Your basal metabolic rate (BMR) is the amount of energy your body needs to function at rest. This accounts for about 60-70% of calories burned in a day and includes the energy required to keep the heart beating, the lungs breathing, the eyelids blinking, and the body temperature stabilized. In general, men have a higher BMR than women. One of the most accurate methods of estimating your basal metabolic rate is the Harris-Benedict formula:

For an adult male:

66 + (6.3 x body weight in pounds) + (12.9 x height in inches) - (6.8 x age in years)

For an adult female:

655 + (4.3 x weight in pounds) + (4.7 x height in inches) - (4.7 x age in years)

The second factor in the equation, physical activity, consumes the next highest number of calories. Physical activity includes everything from making your bed in the morning to jogging. Walking, lifting, bending and just generally moving around burns calories, but the number of calories you burn in any given activity depends on your body weight.

The thermic effect of food is the amount of energy your body uses to digest the food you eat; it takes energy to break food down to its basic elements in order to be used by the body. To calculate the number of calories you expend in this process, multiply the total number of calories you eat in a day by 0.10, or 10%.

⚙ Web Links

How Dieting Works
How Food Works

What causes flatulence?

We all suffer from flatulence to varying degrees, but where does the gas come from?

It happens during the digestive process, which involves breaking things down. Everything in food has to be broken down into small units in order to enter the bloodstream. Protein must be broken into its individual amino acids, fats must be broken into fatty acids, and carbohydrates (both simple and complex) must be broken into individual glucose (or equivalent) molecules.

Flatulence occurs when a food does not break down completely in the stomach and small intestine. As a result, the food makes it into the large intestine in an undigested state. If you are lactose intolerant, for example, you lack the enzyme lactase in your intestine. This enzyme breaks lactose into two sugar molecules so that they can enter the bloodstream. Without lactase, lactose passes undigested through the stomach and small intestine and arrives in the large intestine. There the lactose meets up with trillions of hungry bacteria — the natural "intestinal fauna" we all have in our large intestine. These bacteria are happy to digest lactose. These bacteria produce a variety of gases, in much the same way that yeast produces carbon dioxide to leaven bread. Gases such as methane, hydrogen, and hydrogen sulfide are common gases that these bacteria produce. Hydrogen sulfide is the source of the odor we associate with flatulence.

Certain foods produce more flatulence than others because they contain more indigestible carbohydrates than others. Beans, as you might expect, are particularly well endowed in this regard!

To Your Health ⚙

103

⚙ Web Links

How Food Works

How does your stomach keep from digesting itself?

Your stomach, if you want to be technical about it, is a "crescent-shaped hollow organ" about the size of a large melon. The average adult stomach holds about 3 quarts (3 liters) of fluid. Your stomach is made up of a variety of layers, including:

- **The serosa.** The outer layer that acts as a covering for the other layers.

- **Two muscle layers.** The middle layers that propel food from the stomach into the small intestine.

- **The mucosa.** The inner layer made up of specialized cells, including parietal cells, g-cells, and epithelial cells.

Parietal cells produce hydrochloric acid, a strong acid that helps to break down food. The acid in your stomach is so concentrated that if you were to place a drop on a piece of wood, the acid would eat right through the wood.

The stomach is protected by the epithelial cells, which produce and secrete a bicarbonate-rich solution that coats the mucosa. Bicarbonate is alkaline, a base, and neutralizes the acid secreted by the parietal cells, producing water in the process. This continuous supply of bicarbonate is the main way that your stomach protects itself from *autodigestion* (the stomach digesting itself) and from the overall acidic environment.

In some individuals, due to problems in blood supply to the stomach, or to overproduction of acid, this defense system does not work as well as it should. These people can get gastric ulcers. Also, specific bacteria, called *Helicobacter Pylori*, may cause impairment of the stomach's defenses and can also be responsible for ulcers.

Web Links

How Blood Works
How Cells Work
How Food Works

Who's on First?

Baseball fields often have checkerboards and other patterns mowed into them. How can I create the same effect with my lawn? • In baseball, how does a pitcher throw a curveball? • In scuba diving, what causes "the bends?" • How is the ice in ice-skating rinks made? And how are logos and lines put on the ice? • How is the first-down line superimposed onto the field on tele-vised football games? • Why do golf balls have dimples? • Why does the grass on the greens at a golf course look so perfect? Could my lawn look like this? • How does the ball return work on a coin-operated pool table? • Why shouldn't I go swimming right after I eat? • What makes NASCAR engines different from the engines in street cars?

Baseball fields often have checkerboards and other patterns mowed into them. How can I create the same effect with my lawn?

Groundskeepers at baseball parks have traditionally created checkerboard, diamond, and argyle patterns in the field. These patterns have become more elaborate in recent years. Baseball fans might remember the star patterns created at Coors Field in Denver to commemorate the 1998 All-Star Game. These designs are not all that complicated and can be easily duplicated on your own lawn.

The designs are created through a process called lawn striping. To create lawn-striping designs of your own, you only need two pieces of equipment: a lawnmower and a roller. Many professional groundskeepers use old-fashioned reel mowers to cut a stadium's grass. Attached just behind the blades of the mower is a lawn roller that bends the grass down. Some lawnmower manufacturers are beginning to make riding mowers with full-width rollers mounted to the rear of the mower to make this task easier.

The light shining off the bent grass reveals whatever pattern you make. A checkerboard design is created by passing over the grass in side-by-side rows, first going north to south, then making east-to-west stripes that intersect the north-south stripes. In this way, you alternate the way the grass bends. When you look at your lawn, the stripes of grass leaning away from you will look lighter. This lighter green is caused by the sunlight reflecting off the entire blade of grass. In the darker green stripes, formed by the blades of grass leaning toward you, the sunlight is reflecting only off the tips of the blades.

Different grasses can be used to accentuate the striping effect, including rye grass, fescue, and bluegrass. You won't see such a big contrast between the stripes' colors if you have a warm-season grass, such as Bermuda or zoysia. Watering the grass after mowing can make your pattern stand out even more.

In baseball, how does a pitcher throw a curveball?

A successful major league batter gets a hit only 30% of the time he comes to bat. One of the ways pitchers lower these chances even further is by throwing a curveball. A curveball is a pitch that appears to be moving straight toward home plate but that is actually moving down and to the right or left by several inches. Obviously, a pitch that curves is going to be harder to hit than a fastball that is moving in a straight line.

Two basic factors are involved in creating a curveball:

- Proper grip
- Air flow

Any baseball pitch begins with how the pitcher grips the ball. To throw a curveball, a pitcher must hold the baseball between his thumb and his index and middle fingers, with the middle finger resting on the baseball seam. When the pitcher comes through his motion to throw the ball, he snaps his wrist downward as he releases the ball, which gives the ball topspin. If the pitcher throws properly, the back of his hand will be facing the batter at the end of the motion. The ball will break down and away from a right-handed batter if thrown by a right-handed pitcher.

The spinning action created when the pitcher releases the ball is the secret behind the curveball. This spinning causes air to flow differently over the top of the ball than it does under the ball. The top of the ball is spinning directly into air and the bottom of the ball is spinning with the airflow. The air under the ball is flowing faster than air on top of the ball, creating less pressure, which forces the ball to move down or curve. This imbalance of force is called the *Magnus Effect*, named for physicist Gustav Magnus, who discovered in 1852 that a spinning object traveling through liquid is forced to move sideways.

Adding to the air pressure exerted on the ball are the 108 red stitches that hold the cover on the ball. Because they are raised, the stitches increase the amount of friction created as the air passes around the ball and places more air pressure on top of the ball.

A well-thrown curveball can move as much as 17 inches either way. If you've ever seen a batter jump out of the way of a baseball that ends up crossing over the plate, you've seen a good curveball.

In scuba diving, what causes "the bends?"

When using self-contained underwater breathing apparatus (SCUBA, or scuba) equipment, a diver breathes highly compressed air from a tank.

Scuba diving is different from holding your breath and diving, and to understand the difference you need to understand the incredible pressures that a diver's body experiences. When we are living on dry land at sea level, the air around us has a pressure of 14.7 pounds per square inch (PSI). That is a "normal pressure" for our bodies. Because water is so heavy compared to air, it does not take much water to exert a lot of pressure. For example, a 1" x 1" column of water 33 feet high weighs 14.7 pounds, so it exerts another 14.7 PSI.

If you hold your breath and dive down 33 feet, therefore, your lungs actually contract in size by a factor of 2. They have to — there is twice as much pressure around the air in your lungs, so they contract. When you rise back up, the air expands again so your lungs return to normal size.

When you breath from a scuba tank, the air coming out of the tank actually has the same pressure as the pressure the water is exerting. It has to, or it won't come out of the tank. Therefore, when scuba diving, the air in your lungs at a 33-foot depth has twice the pressure of air on land. At 66 feet, it has three times the pressure. At 99 feet, it has four times the pressure, and so on.

When high-pressure gases in the air come in contact with water, they dissolve into the water. This is how carbonated beverages are made. To make carbonated water, water is exposed to high-pressure carbon dioxide gas and the gas dissolves into the water. We all know what happens when you release the pressure in a bottle of soda: Bubbles suddenly start rising. The gas dissolved in the water at high pressure comes out of the liquid when the pressure is released, and we see it as bubbles.

If a scuba diver stays under water, say at a depth of 100 feet, for a period of time, some amount of nitrogen from the air will dissolve in the water in his/her body. If the diver were to swim quickly to the surface, the effect on his body is the same as it is when you uncork a bottle of soda: The gas is released. This release can cause a very painful — sometimes fatal — condition called decompression sickness or "the bends."

To avoid the effects of quick decompression, the diver must rise slowly so that the gas can come out of his blood and tissue slowly. If the diver does rise too fast, the only cure is either to go back down (which is like putting the cap back on a bottle of soda — the bubbles stop), or to enter a pressurized chamber in which the air pressure matches the pressure at depth. Then the pressure is released slowly.

Decompression sickness is one danger of diving. Other dangers include nitrogen narcosis, oxygen toxicity, and simple drowning (running out of air before making it back to the surface). If the diver decompresses properly, remains at "recreational depths" (less than 100 feet or so), and is careful about the air supply, the dangers can be largely eliminated. Proper training, good equipment, and careful execution are the keys to safe diving.

How is the ice in ice-skating rinks made? And how are logos and lines put on the ice?

Occasionally on college campuses across the northern U.S., you can see an incredible display of ingenuity during the winter months: Students have been known to make skating rinks on their front lawns! To make the rink, they hose down the snow on the lawn and let that freeze to get a hard surface. Then they lay water on top in thin layers so that it can freeze quickly, until it is smooth. The result is an instant skating rink! The cold air and ground temperatures are enough to freeze the water and keep the ice solid for several months. Then in the spring the whole thing melts and the students get their lawn back.

As you might guess, this process doesn't work so well in Florida. Indoor skating rinks almost always use cold concrete to make the ice. When the rink is built, miles of metal pipes are laid inside a concrete slab. A large refrigeration plant produces ice-cold glycol that runs through these pipes. The entire slab of concrete that makes up the arena floor drops below freezing. Then thin layers of water are poured on the concrete and allowed to freeze.

As for the lines and logos, they are literally painted onto the ice. A water-based paint is used to cover the ice surface and paint the decorations. After the first layer of ice is put out, the entire surface is painted one solid color. Then, after another layer of ice is set, the lines and logos are painted. This is usually done by hand with stencils. Then more ice is added on top of the paint to protect it.

How is the first-down line superimposed onto the field on televised football games?

This concept is one of those things that sounds really simple in theory, but then ends up being incredibly complicated when you actually try to do it. The system that ESPN uses to paint the line is called 1st and Ten and is created by a company called SporTVision.

The simplest description of this system is as follows: Knowing where the first-down line is (for example, at the 42-yard line), use a computer to draw that line on the field so that TV viewers can see the line as though it were painted on the field.

Here are some of the problems that have to be solved in order for this system to work:

- The system has to know the orientation of the field with respect to the camera so that it can paint the first-down line with the correct perspective from that camera's point of view.
- The system has to know, in that same perspective framework, the exact location of every yard line.
- Given that the cameraperson can move the camera, the system has to be able to sense the camera's movement (tilt, pan, zoom, focus) and understand the perspective change as a result of the movement.

- Given that the camera can pan while viewing the field, the system has to be able to recalculate the perspective at a rate of 30 frames per second as the camera moves.

- A football field is not flat; it crests very gently in the middle to help rainwater run off. The line calculated by the system has to appropriately follow the curve of the field.

- A football game is shot by multiple cameras at different places in the stadium, so the system has to do all of this work for several cameras.

- The system has to be able to sense when players, referees, or the ball cross over the first-down line so that it does not paint the line over top of them.

- The system has to be aware of superimposed graphics that the network might overlay on the scene.

There are probably several other glitches as well. Painting that simple line on the field is quite a challenge!

To solve these problems, the creators of the 1st and Ten system combine computer hardware and software. First, each camera must have a very sensitive encoder attached to it that can read the camera's angle, tilt, zoom, and so on and send that information to the system. The system must also have a detailed 3-D model of the field so that it knows the location of each yard line. By integrating the tilt, pan, and zoom information with the 3-D model, the system can begin to calculate where the line should go. Then the system uses color palettes for the field and the players/referees/ball to recognize, pixel by pixel, whether it is looking at the field or something else. This way, only the field gets painted.

All of this computation requires a lot of equipment: eight computers, several camera encoders, and a lot of wiring running from the cameras to a truck that holds all the equipment.

○ Web Links

How the First-Down Line Works

Why do golf balls have dimples?

The reason golf balls have dimples starts with natural selection. Originally golf balls were smooth, but golfers noticed that older balls that were beat up with nicks, bumps, and slices in the cover seemed to fly farther. Golfers, being golfers, naturally gravitate toward anything that gives them advantage on the golf course, so old, beat-up balls became standard issue.

At some point an aerodynamicist must have looked at this problem and realized that the nicks and cuts were acting as turbulators — they induce turbulence in the layer of air next to the ball (the "boundary layer"). In some situations a turbulent boundary layer will reduce drag.

There are two types of flow around an object: laminar and turbulent. Laminar flow has less drag, but it is also prone to a phenomenon called *separation*. Once separation of a laminar boundary layer occurs, drag rises dramatically because of eddies that form in the gap. A turbulent boundary layer has more drag initially but also better adhesion, and therefore is less prone to separation. Therefore, if the shape of an object is such that separation occurs easily, it is better to turbulate the boundary layer (at a slight cost of increased drag) in order to increase adhesion and reduce eddies (a significant reduction in drag). Dimples on golf balls turbulate the boundary layer. The dimples on a golf ball are simply a formal, symmetrical way of creating the same turbulence in the boundary layer that nicks and cuts do.

Why does the grass on the greens at a golf course look so perfect? Could my lawn look like this?

The grass on a well-maintained golf green is absolutely amazing — it is a flawless surface made out of plants! To make the lawn this perfect takes a lot of work...

First, you need to decide where you put the green. It needs to be a spot that provides plenty of sunlight (preferably full sunlight with no surrounding trees) and good airflow over the grass.

After you choose the location, the physical work starts by creating what is practically a hydroponic, or soilless, system for growing the grass. When constructing the green, a bulldozer creates a 12- to 16-inch (30- to 40-centimeter) deep hole the size of the green. In the most advanced systems, this hole is completely lined with plastic, and then gravel, drainage pipes, and sand are added. The green's grass grows in a sterile sand medium with perfect drainage. The surface is contoured to allow perfect run-off as well, so there is no puddling when it rains.

Next you choose perfect grass. Two of the most popular types of turf are creeping bentgrass and bermuda.

A sterile sand medium and a good location controls for a huge number of variables, but now the grass is totally dependent on its keepers for life support. That means the grass needs a steady diet of water and nutrients to keep it alive. To this mix is added a variety of herbicides (to kill weeds that try to move in), pesticides (to control insect damage), and fungicides (to control disease) to help keep the grass perfect.

After the green is established, you start in on maintenance. This includes daily mowing with a precision green mower, watering, fertilizing, applying chemicals, aerating, and general coddling.

How does the ball return work on a coin-operated pool table?

If you have ever played one of those coin-operated pool tables in a restaurant or arcade, then you know that the obvious question is, "how the heck does this thing return the cue ball to me???"

If you look inside the table, you would see a system of chutes that connect to the six pockets on the table. Each chute sends a pocketed ball from the pocket to a collection chamber, where the numbered balls are lined up single file. These numbered balls remain locked in the chamber, which you can see behind a piece of Plexiglas, until someone inserts some coins to play a game. Of course the cue ball can't get stuck in this chamber — if a player accidentally pockets the cue ball (a scratch), the cue ball needs to come back out.

Table manufacturers needed a way to allow the cue ball to be returned to play, while keeping the pocketed numbered balls locked in the storage compartment. Many solutions have been proposed to solve this problem, including:

- Light sensors that sense the light reflected from the cue ball.
- A metallic core ball that would trigger a separating mechanism when it passed through an electromagnetic field.
- A balance mechanism that would separate a heavier cue ball from lighter numbered balls.

And so on. For the most part, though, coin-operated tables use two types of cue balls that can be easily separated:

- An oversized ball that is separated by a radius-gauging device.
- A magnetic cue ball that triggers a magnetic detector.

The oversized ball is approximately 2 3/8 inches (6 centimeters) in diameter, which is about 1/8 inch (2 millimeters) larger than a normal ball. This slight difference in size allows the cue ball to be separated before it gets to the storage compartment. The smaller numbered balls are able to pass through a gauging mechanism, while the larger cue ball is directed through a second chute, where it falls out into an opening on the side of the table.

For players who dislike using the slightly larger cue ball, they can play on the coin-operated machines that use a magnetic ball — a magnet is built into the core of the cue ball. Magnetic cue balls that go into a pocket are separated from numbered balls by a magnetic detector. As the magnetic ball passes this detector, the magnet triggers a deflecting device that separates the cue ball and, again, sends it into the opening on the side of the table.

Both the oversized and magnetic cue balls can be used interchangeably on most of today's coin-operated tables, but each has its shortcomings. If you are a beginning pool player the larger ball might not affect your play, but it can disrupt the play of some advanced players who are used to playing with the normal 2 1/4-inch cue ball. Likewise, some players will notice a difference in the properties of a magnetic ball, which sometimes lacks a true roll. Also, because the magnetic ball has the magnetic material inserted into it, it has a greater tendency to shatter if dropped on a hard surface.

Why shouldn't I go swimming right after I eat?

"Don't go swimming for an hour after you eat" is a good piece of advice. If you do hop into the pool or the ocean right after you eat, you could develop cramps and risk drowning.

The key to understanding the risk is knowing that your body will always work to take care of its energy needs. Conflicting needs can cause problems. When you exercise, your sympathetic nervous system, a part of the automatic or autonomic nervous system (brainstem, spinal cord) stimulates the nerves to your heart and blood vessels. This nervous stimulation causes those blood vessels to contract or constrict. This constriction increases the resistance of the blood vessels in those tissues and reduces blood flow to those tissues. Working muscle also receives the command for its blood vessels to constrict, but the metabolic byproducts produced within the muscle override this command and cause the blood vessels to dilate. So if most of your body is getting the message to cut the blood flow and your muscles are getting the message to boost the blood flow, the blood that would have gone to some of your organs will go instead to your muscles. Your body is taking from one part to give to another part, but it's okay if the organs that are getting less blood, such as your stomach or your kidneys, are not working.

But what if one of those organs does need the blood to do its work? If you have just eaten, the food in your stomach begins to be digested. This requires a greater blood supply to the stomach and intestines. Like metabolic byproducts in working muscle, the presence of food in the stomach overrides the commands by the nervous system to constrict the blood vessels in the stomach and intestines. Now you have a situation where the digestive system and working muscle both have increased demands for blood flow and are competing for the increased blood supply. As a result, neither system gets enough blood flow to meet its needs and the tissues begin to cramp. If you are in water, this cramping presents a serious problem and increases your risk of drowning. If you wait about an hour to allow some digestion to occur and food to leave your stomach, your risk of cramps goes down.

What makes NASCAR engines different from the engines in street cars?

We visited Bill Davis Racing in High Point, North Carolina, to answer this question. What we learned was surprising — these engines actually have quite a lot in common with street car engines.

Bill Davis Racing runs two NASCAR teams, the Caterpillar-sponsored No. 22 car and the Amoco-sponsored No. 93 car. In 2001 both of these teams raced Dodge Intrepid cars.

Dodge provides the engine block and cylinder head for the engine. They are based on a 340-cubic inch V-8 engine design that was produced in the 1960s. The actual engine blocks and heads are not made from the original tooling; they are custom-made race engine blocks, but they do have some things in common with the original engines. They have the same cylinder bore centerlines, the same number of cylinders, and the same base displacement. And like the original 1960s engines, the valves are driven by pushrods.

The engines in today's NASCAR race cars produce upward of 750 horsepower, and they do it without turbochargers, superchargers, or particularly exotic components. Here are some of the factors involved in creating so much power:

- The displacement is large: 358 cubic inches (5.87 liters). Not many cars have engines this big, but the ones that do usually generate well over 300 horsepower.
- The NASCAR engines have extremely radical cam profiles. These profiles open the intake valves much earlier and keep them open longer than street cars. This allows more air to be packed into the cylinders, especially at high speeds.
- The intake and exhaust are tuned and tested to provide a boost at certain engine speeds. They are also designed to have very low restriction, and there are no mufflers or catalytic converters to slow down the exhaust.
- They have carburetors that can let in huge volumes of air and fuel. There are no fuel injectors on these engines.
- They have high-intensity programmable ignition systems so

that the spark timing can be customized to provide the most possible power.

- All of the subsystems such as coolant pumps, oil pumps, steering pumps, and alternators are designed to run at sustained high speeds and temperatures.

When these engines are assembled, they are built to very exacting tolerances (parts are machined more accurately), so that everything fits perfectly. Cylinders are bored to more exacting tolerances than they are in street cars. The crankshafts and other rotating parts are balanced. Making sure that the parts are as close to their exact dimensions as possible helps the engine achieve its maximum potential power and also helps reduce wear. If parts are too big or small, power can be lost due to extra friction or pressure leakage through bigger than necessary gaps.

After the engine is assembled, it is broken in by running on the dynamometer, which measures engine power output, for 30 minutes. The engine is then inspected. The filters are checked for excess metal shavings to make sure no abnormal wear is taking place.

If the engine passes these tests, it goes on the dynamometer for another 2 hours. During this test, the ignition timing is dialed in to maximize power and the engine is cycled through various speed and power ranges.

Next, the engine is inspected thoroughly. The valve train is pulled and the camshaft and lifters are inspected. The insides of the cylinders are examined with borescopes, which use mirrors to inspect the interior. The cylinders are pressurized and the rate of leak down (rate at which the pressure drops) is measured to see how well the pistons and seals hold the pressure. All of the lines and hoses are checked.

Only after all of these tests and inspections are finished is the engine ready to go to the races. Insuring the reliability of the engine is critical — almost any engine failure during a race eliminates any chances of winning.

⚙ Web Links

How Car Engines Work

Cool Problems/ Cool Solutions

⚙ What would happen if I drilled a tunnel to the center of the earth and jumped into it? • I often hear the expression, "If I had all the money in the world…" So, how much is all the money in the world? • Can a telescope detect the equipment left behind by astronauts on the moon to prove/disprove missions? • How much ice would I have to store up in the winter if I wanted to air condition my house all summer? • How many sheets of paper can be produced from a single tree? • If you took all the matter in the universe and pushed it into one corner, how much space would it take up? • Is there a way to actually see a satellite in orbit? • If daytime running lights were mandatory in the United States and all vehicles had them, how much extra gasoline would be used each year? • If you could build a train that could travel as fast as a bullet, what would happen if you fired a gun from the back or from the front of the train? • How much does the earth weigh?

What would happen if I drilled a tunnel to the center of the earth and jumped into it?

Although it would be impossible to do this on earth, you actually could do this on the moon. The moon has a cold core and it also doesn't have any oceans or groundwater to complicate things. In addition, the moon has no atmosphere, so the tunnel would have a nice vacuum in it that eliminates aerodynamic drag.

So, imagine that you drilled a tunnel through the moon. Imagine that, down one side, is a ladder. If you were to climb down the ladder, you would find that your weight decreases. Because gravity is caused by objects attracting one another with their mass, as you descend into the tunnel, more and more of the moon's mass is above you, so it attracts you upward. Once you climb down to the center of the moon, you would be weightless. The mass of the moon is all around you and attracting you equally, so it all cancels out and you would be weightless.

If you were to actually leap into the tunnel on one side of the moon and allow yourself to fall, you would accelerate toward the center at a very high speed. Then you would zoom through the center and start decelerating. You would eventually stop when you reached the tunnel's lip on the other side of the moon, and then you would start falling back down the tunnel in the other direction. You could oscillate back and forth like this forever, or you could grab the lip on the other side.

If you could do this on earth, one amazing effect would be the ease of travel. The diameter of the earth is about 7,800 miles (12,700 kilometers). If you drilled the tunnel straight through the center and could create a vacuum inside, anything you dropped into the tunnel would reach the other side of the planet in just 42 minutes! People could travel from the U.S. to China or India in less than an hour, for free!

I often hear the expression, "If I had all the money in the world..." So, how much is all the money in the world?

To make this question answerable in a finite amount of time, let's make a simplification and ask, "How much money is there in the United States?" Because the statistics for the U.S. are easy to come by, we can look at this question in a couple different ways.

The first way to look at it might be, "How much cash is there in U.S. currency?" In other words, if you took all the bills and coins floating around today, how much money is that? According to the Federal Reserve, the current number is very close to half a trillion dollars! That sounds like an incredible amount, but think about it this way: In 1990 there were about 250 million people in the U.S. If you took all the cash and divided it equally, each person would only get about $2,000. Obviously more money than that is floating around.

The rest of the money is held in various types of bank accounts, and the Federal Reserve tracks these funds in three different values: The M1, M2, and M3 money supplies.

M1 is all the currency, plus all the money held in checking accounts and other checkable accounts, as well as all the money in travelers' checks. As of September 1999, the M1 money supply was approximately $1.1 trillion.

M2 is M1 plus all the money held in money market funds, savings accounts, and small CDs. As of September 1999, the M2 money supply was approximately $4.6 trillion.

M3 is M2 plus all of the large CDs. As of September 1999, M3 was approximately $6.25 trillion.

So if you wanted all of the money in the United States, what you would ask for is M3, and you would get approximately $6 trillion. How much money is that? It's a lot, but if you look at the federal budget you will see it is only enough money to run the U.S. federal government for about 3 years.

Can a telescope detect the equipment left behind by astronauts on the moon to prove/disprove missions?

In order to answer this question, you need to understand something about the resolving power of telescopes. A typical orbiting telescope pointed at earth has 1-meter (3.3-feet) resolution, which means that something on the ground that is 1 meter square (10.8 square feet) produces 1 pixel in the image. In an image of the Statue of Liberty, her head is about 3 meters (10 feet) across, so it should take up about 3 x 3 pixels in the image.

The best telescope available today is the Hubble Space Telescope. The Hubble telescope would have approximately 15-centimeter (6-foot) resolving power if it were pointed at something on the earth like the Statue of Liberty. The Statue of Liberty's head would fill perhaps 18 x 18 pixels in an image taken by the Hubble Space Telescope.

The moon is about 1,000 times farther away from the Hubble Space Telescope than the earth is. That means that if you pointed the Hubble at the moon, it would have 150-meter (492-feet) resolution. At that resolution, a football stadium occupies just 1 or 2 pixels of the image. That means that there would be no way to discern the Lunar Excursion Module or any of the other equipment left on the moon. It is just too small to pick up, even with the world's best telescope.

Even though you cannot see direct evidence of the moon excursions by using a telescope, there is one artifact that the astronauts left behind on the moon that does provide evidence of their missions. That artifact is a laser beam reflector that has been used to track the distance of the moon from the earth.

A true skeptic could argue that this reflector was placed on the moon by a robotic satellite. Its presence does prove that humans have been able to put something on the moon through some means, and curbs critics who claim that all NASA missions of any type are a hoax.

How much ice would I have to store up in the winter if I wanted to air condition my house all summer?

You could easily build an ice—powered air conditioner. All you need is a big insulated container (probably in the form of a hole in the ground) with some coiled tubes at the bottom. You would run a chilled water circuit from the container to a radiator inside the air conditioner. In addition, you would need a small pump to pump the water in the chilled water loop.

Let's make a couple of assumptions:

- Let's assume that your air conditioner runs for 12 hours a day for 3 months of the year.

- Let's assume that your house has a 5-ton air conditioner (60,000 British Thermal Units, or BTUs).

- Let's assume that you can store the snow and ice with 50% efficiency. That is, over the course of the summer you will lose half of the ice to melting, inefficiencies in the system, and other causes.

To cool the house you therefore need:

60,000 BTUs/hr x 12 hours/day x 90 days= 64,800,000 BTUs

Multiplying by our 50% efficiency rating, let's call it 130 million BTUs.

If you have a gram of ice at 0°C (32°F), it will absorb 80 calories of energy converting from ice to liquid water. There are 252 calories in a BTU. So we need 3.15 grams of water to absorb one BTU of heat.

So we need: 130,000,000 BTUs x 3.15 grams/BTU= 409,500,000 grams of ice. That's about 410,000 liters of ice, or 410,000 kg (902,000 pounds) of ice that you must store to cool your house all summer. That's a cube 740 centimeters (24.26 feet) on a side. Very roughly speaking, you would have to dig a hole as big as your house and insulate it well, and then in the winter you would have to shovel it full of nearly a million pounds of ice. But if you do that, you can cool your house for free! (The value of the equivalent electricity to cool the house, at 7.5 cents per kilowatt-hour, would be about $1,500.)

How Air Conditioners Work

How many sheets of paper can be produced from a single tree?

Although it is difficult to give an exact number of sheets, here is how you would guess an answer to this question. First you need to define what a "tree" is. Is it a giant redwood or a little weeping willow?

Most paper is made from pine trees. Most mature pines are about a foot (0.3 meters) in diameter and 60 feet (18.3 meters) tall. Ignoring taper, that's about 81,430 cubic inches (1.3 cubic meters) of wood ($\pi \times radius$ squared $\times length$ = 3.14 × 6^2 × 60 × 12). A 2 × 4 piece of lumber weighs about 10 pounds (4.5 kilograms) and contains 504 cubic inches (0.008 cubic meters) of wood (3.5 × 1.5 × 8 × 12). That means a pine tree weighs roughly 1,610 pounds (81,430 / 504 × 10), or approximately 730 kilograms. In paper manufacturing, you turn the wood into pulp. The yield is about 50%—about half of the tree is knots, lignin, and other materials that are no good for paper. So a pine tree yields about 805 pounds (365 kilograms) of paper.

A ream of paper weighs about 5 pounds (2.27 kilograms) and contains 500 sheets. (You often see paper described as "20 pound stock" or "24 pound stock"; this description refers to the weight of 500 sheets of 17" × 22" paper.) So a tree would produce roughly 805 / 5 × 500, or 80,500 sheets of paper.

If you took all the matter in the universe and pushed it into one corner, how much space would it take up?

This question does contain some unknowns. If you are willing to accept three assumptions, however, we can come up with a reasonable answer.

The first question is, "How big is the universe?" No one knows, but let's assume that the universe is a cube that is 30 billion light years on each side. That means that the total universe contains about 2.7×10^{31} cubic light years.

The next question is, "How much matter does the universe contain?" The mass of the universe is a source of debate right now because there is no easy way to put the universe on a scale. Scientists use different techniques to estimate the mass. One technique comes up with an estimate of about 1.6×10^{60}) kilograms (3.53×10^{60} pounds) for the mass of the universe. Other estimates give other numbers, but all are in that same basic ballpark.

The next question is, "What density do you want to assume the mass will have once you push all of it into one corner?" If you were really to do this — if you actually did move all of the mass of the universe into one corner — it would condense instantly into a black hole and potentially ignite another big bang. Let's say that you could keep it from exploding, and you were somehow able to keep all of the mass evenly distributed at the density of the sun. The density of the sun is about 1,410 kilograms (3,102 pounds) per cubic meter. (For comparison, the density of water is 1,000 kilograms or 2,205 pounds per cubic meter.)

If you are willing to accept these three assumptions, then:

> 1.6×10^{60} kilograms / 1,410 kilograms per cubic meter = 1.1×10^{57} cubic meters of matter in the universe

A cubic light year contains about 1×10^{48} cubic meters. So all of the matter in the universe would fit into about 1 billion cubic light years, or a cube about 1,000 light years on each side. That means that only about 0.0000000000000000000042% of the universe contains any matter. The universe is a pretty empty place!

Is there any way to actually see a satellite in orbit?

If you or a neighbor has a satellite dish sitting in the yard to pick up television signals, then you know where at least one orbiting satellite is located; the dish is pointed right at it! TV satellites, unfortunately, live in geosynchronous orbits approximately 22,000 miles (35,420 kilometers) away, so it is impossible to see them unless you have a very big telescope.

However, lots of satellites pass overhead in asynchronous orbits that are only 200 or 300 miles (321 to 482 kilometers) away. The space station is the biggest satellite of all and is very easy to pick out if you know its schedule. If you live in a place that has a very clear sky (where you can still see the Milky Way, for example), simply lie on your back on a moonless night and look carefully. Occasionally you will see something that looks like a star, but that is moving. That's a satellite! This technique works especially well on a boat in the Caribbean, close to the equator.

> **Other hints for viewing satellites:**
>
> Satellite tracking software is available for predicting orbit passes. You should note the exact times satellites are expected to pass.
>
> Use binoculars on a clear night when the moon is not bright.
>
> Ensure your watch is set to match exactly a known time standard, so that your timing will not be off; you wouldn't want to miss something because you have the wrong time!

 **Web Links**

How Satellites Work

If daytime running lights were mandatory in the United States and all vehicles had them, how much extra gasoline would be used each year?

For several years now Canada has required all new cars to have daytime running lights. Any time the car is running the head-lights are on, but the taillights and other lights are off. The driver has to turn on these other lights from the dashboard at night. Studies seem to indicate that having the headlights on during daylight hours reduces the number of multiple vehicle accidents. Conversely, there has been some controversy about the negative effects of these lights in which people forget to turn on the other lights at night — a mistake that causes extra accidents, and a good example of the "law of unintended consequences."

The U.S. has not adopted this law, but if it did the law would definitely cause an increase in gasoline consumption. Headlights require power, and a car's engine produces power by using gaso-line. If you make a few assumptions, it is possible to estimate how much gas the law would consume.

A typical headlight bulb uses about 55 watts. Sometimes the day-time running lights run at a lower wattage so they use a little less power. Let's say the daytime running lights use 100 watts since there are two bulbs.

To calculate the energy used, we need to figure out how much time drivers will spend with their lights on. According to the National Highway Traffic Safety Administration (NHTSA), vehicles in the U.S. drove 2,560 billion miles in 1997. We need to make a guess at the average speed people drive, including stops, in order to figure out how much time people spent driving their cars. Let's guess 30 mph, which means each mile takes 2 minutes. That makes 5,120 billion minutes or 85.3 billion hours. If each car normally drives at night about half the time, that means that the daytime running lights would be on 42.6 billion hours a year. Multiplying by the 100 watts we get 4,260 billion watt-hours or 4.26 billion kilowatt-hours. The U.S. uses about that much electricity nationwide in 12 hours.

Now we need to figure out how much electrical energy we can get out of a gallon of gas. A gallon of gas contains about 60 kilowatt-hours of chemical energy, but this energy has to go through two conversion processes before it can be used in a light bulb. First, the car's engine must turn the chemical energy into mechanical power. Car engines don't do this very efficiently; only about 25% of the chemical energy can be turned into mechanical power, and the rest is wasted as heat. After the engine gets done with our gallon of gas, we have 15 kilowatt-hours left.

Now the alternator on the car has to turn the mechanical power from the engine into electrical power. The alternator does this a lot better than the engine, but it is still only about 70% efficient. In the end, we get about 10.5 kilowatt-hours of electrical energy out of a gallon of gas.

To calculate how many gallons of gas this is, you can divide the 4.26 billion kilowatt hours of energy that the daytime running lights consume each year by the 10.5 kilowatt-hours of energy each gallon of gas yields. If daytime running lights were on all the vehicles in the U.S., we would burn an extra 406 million gallons of gas each year. That's only a couple gallons for each vehicle, but in total it is more than all of the vehicles in the country burn in a day. At $1.50 a gallon, that's $600 million per year. Looking at it another way, an extra 8 billion pounds (3.6 billion kilograms) of carbon dioxide would be added to the atmosphere by this law.

It's an interesting question because it shows how a simple idea like, "Let's have everyone turn on their headlights all the time" can carry a real cost when you try to implement it. Whether the benefit is worth the cost is an important question in almost any public policy decision.

If you could build a train that could travel as fast as a bullet, what would happen if you fired a gun from the back or from the front of the train?

This is a good question because it involves the concept of reference frames. The bullet will always travel at the same speed. In other reference frames, however, unexpected things can happen!

Imagine you are on a perfectly smooth speeding train, moving at a uniform speed (not accelerating or turning), in a car with no windows. You would have no way of knowing how fast you are going (or if you are moving at all). If you throw a ball straight up in the air, it will come straight back down whether the train is sitting still or going 1,000 mph. Because you and the ball are already moving at the same speed as the train, the only forces acting on the ball are your hand and gravity. So the ball behaves exactly as it would if you were standing on the ground and not moving.

So what does this mean for a gun? If the gun shoots bullets at 1,000 mph (1,609 kph), then the bullet will always move away from the gun at 1,000 mph. If you go to the front of a train that is moving at 1,000 mph and shoot the gun forward, the bullet will move away from you and the train at 1,000 mph, just as it would if the train were stopped. But, relative to the ground, the bullet will travel at 2,000 mph (3,219 kph), the speed of the bullet plus the speed of the train. So if the bullet hits something on the ground, it would hit it going 2,000 mph.

If you shoot the bullet off the back of the train, the bullet will still be moving away from you and the gun at 1,000 mph, but now the speed of the train will subtract from the speed of the bullet. Relative to the ground, the bullet will not be moving at all, and it will drop straight to the ground.

What's true for bullets, however, is not true of some other things that you might "shoot" from the front of the train. A great example is sound waves. If you turn on the stereo in your living room, sound waves "shoot out" of the speaker at the speed of sound — something like 700 mph (1,127 kph). The waves propagate through the air at that fixed speed, and they can go no faster. If you put a speaker at the front of the 1,000 mph train, the sound waves will not depart the train at 1,700 mph (2,736 kph). They cannot go faster than the speed of sound. This is the reason why planes traveling faster than the speed of sound create sonic booms.

⚙ Web Links

How Maglev Trains Will Work

How much does the earth weigh?

The quick answer is: approximately 6,000,000,000,000,000,000,000,000 (6×10^{24}) kilograms.

The interesting sub-question is, "How did anyone figure that out?" It's not like the planet steps onto the scale each morning before it takes a shower. It turns out that with a large object like a planet you can figure out its weight by calculating its gravitational attraction for other objects.

Any two masses have a gravitational attraction for one another. If you put two bowling balls near each other, they will attract one another gravitationally. The attraction is extremely slight, but if your instruments are sensitive enough you can measure the gravitational attraction that two bowling balls have on one another. From that measurement you can determine the mass of the two objects. The same is true for two golf balls, but the attraction is even slighter because the amount of gravitational force depends on the amount of mass in the objects, and golf balls are obviously smaller than bowling balls.

So let's look at the actual calculation—it is a little mathy, but the technique is cool. Newton showed that, for spherical objects, you can make the simplifying assumption that all of the object's mass is concentrated at the center of the sphere. The following equation expresses the gravitational attraction that two spherical objects have on one another:

$$F = G \times M1 \times 2/R^2$$

R is the distance separating the two objects, in meters. G is a constant that is $6.67259 \times 10^{-11} m^3/s^2 kg$. $M1$ and $M2$ are the masses of the two objects that are attracting each other. F is the force of attraction between them.

Assume that earth is one of the masses ($M1$) and a 1-kg sphere is the other ($M2$). $M1$ is the "unknown" in the equation. We can calculate the force easily enough, simply by dropping the 1-kg sphere and measuring the acceleration that the earth's gravitational field applies to it. The force between them is 9.8 kg $\times$ m/s^2. The radius of the earth is 6,400,000 meters. If you plug all of these values in and solve for $M1$, you find that the mass of the earth is 6,000,000,000,000,000,000,000,000 (6×10^{24}) kilograms.

It is more proper to talk about mass rather than weight because weight is a force that requires a gravitational field. You can take a bowling ball and weigh it on the earth and on the moon. The weight on the moon will be 1/6 that on the earth, but the amount of mass is the same in both places. The mass of the earth is a constant.

6,000,000,000,000,000,000,000,000 (6×10^{24}) is a lot of mass. It's so much that it is hard to get a handle on this measurement. Here are three comparisons that might help you get a perspective:

- There are roughly 6 billion people on the planet. If you assume that each person weighs 70 kg, then all the people have a mass of 420,000,000,000 kg. This mass hardly makes a dent in the earth's total mass.

- There are approximately 1,000,000,000,000,000,000,000 (1×10^{21}) kg (2.2×10^{21}) of water on earth in the form of oceans, lakes, ice caps, and vapor in the atmosphere. That means that about 0.02% of the earth's mass is water.

- The mass of the sun is about 2×10^{30} kg. So the sun is equal to about 330,000 earths.

A good portion of the earth's weight is made up of its huge iron core.

INDEX

A

AC. *See* Alternating current
Acid rain, 68
Acoustic finish of ceiling, 73
Active sensors, 29
Active Server Pages (ASP), 36
Agar, 85
Aioli, 91
Air conditioner,
    ice-powered, 123–24
Aircraft search light, 23
Air flow of the baseball, 107
Air intake problem, 54–55
Airplane
  locks, 63–64
  magneto used on, 64
  octane ratings of gasoline in, 57
  oxygen canisters on, 62
  producing sound waves, 25
  supersonic speeds, flying at, 26
Alkaline, 104
Alka-Seltzer, 17
Alphaltic bitumen, 47
Alternating current (AC), 32
Alternator on car, 128
Aluminum, 13
Amber color of mood ring, 16
Amino acids, 103
Amplifier, feeback in PA
    system and, 30–31
AMPS. *See* Analog cell phones
Analog cell phones (AMPS), 76
Animals
  birds of preys, vision of, 96
  dogs, chocolate and, 86
  shedding, 94
  zebra's stripes, 24–25
Animated GIF files, 40
Appliances. *See* Home appliances
Arabica plant, 88
Arcade games, light guns on, 4–5
Armature of magneto, 64–65
ASP. *See* Active Server Pages
.asp extension, 35–36
Asynchronous orbits of
    satellites, 126
Atomic clocks, 79–80
Autodigestion, 104

Automatic door openers, 29
Automobiles. *See also* Engines
  alternator, 128
  daytime running lights, 127–28
  measuring speed of, 65–66
  rearview mirror of, 50
Autonomic nervous system, 115

B

Baking powder reaction, 17
Baking soda, 17
Balance, sense of, 97
Basal metabolic rate (BMR), 102
Baseball
  checkerboard designs on
    fields, 106
  curveballs, 107–8
  red stitches of, 107
BCD. *See* Binary Coded Decimal
Beer
  alcoholic versus nonalcoholic, 83
  making, 84
  root beer, 83–84
  widget inside can, 87–88
Bends, the, 108–9
Bentgrass, 113
Bermuda grass, 106, 113
Beverages, carbonated, 108. *See
    also* Soft drinks
Bicarbonate, 104
Big bang, 125
Bill Davis Racing, 116
Binary Coded Decimal (BCD), 79
Birch beer, 83
Birds of prey, vision of, 96
Birthday candles, 12
Bitmap images, 40
Black-and-white movies, coloriza-
    tion of, 18–19
Black color of mood ring, 16
Black hole, 125
Bladder, 98
Blindness, legal, 96
Blue color of mood ring, 16
Bluegrass, 106
Blue-green color of mood ring, 16
Blue light, 12
BMR. *See* Basal metabolic rate
Body
  hair, 94
  surface temperature, 15–16

weight, as factor in calculating caloric need per day, 102
Boiling point of water, 14
Borescopes, 117
Boundary layer of golf balls, 112
Brain, human's, 42
Brakes, 48
BTU, 123–24
Bulb
  incandescent, 32
  light, amount of coal needed to power, 68
  light, electricity consumption of, 62
Bullet, silencing, 27–28
Burglar alarm with motion sensors, 30
Butane, 56
Byte, in caller ID, 77

C

Caffeine
  removing from coffee beans, 88–89
  toxic levels of, 86
Calculations
  basal metabolic rate, 102
  caloric need per day, 102–3
  gravitational attraction of two spherical objects, 130
  light year, 26
  kilowatt hours, 39
Caller ID on telephone, 76, 77
Calories
  body consumption of, 102–3
  count in soft drink, 83
Camera rotating around frozen actor in movies, 5–6
Camouflage of zebra's stripes, 24–25
Candles, trick birthday, 12
Candy
  Pop Rocks, 18
  Wint-O-Green Lifesavers, 12
Canisters, oxygen, 61–62
Capacitance, 78
Carbohydrates, explosiveness of, 61
Carbon atoms, chain lengths of, 46–47, 56
Carbon dioxide
  in beer, 87
  frozen. See Dry ice
  gas, in Pop Rocks candy, 18
  as pollutant, 68
  in Pop Rocks candy, 18
  processing of caffeine, 88–89
Carbonated beverages, 108. See also Soft drinks
Carbonless copy paper, 17
Carrageenan, 85–86
Cars. See Automobiles; Engines
Catalase, 95
Cavitation, 99
CB radios, 9
CCA. See Chromated copper arsenate
Ceiling, textured finish on, 73
Cell phones, analog versus digital, 76
Cellulose gum, 85
Central air conditioner, electricity consumption of, 62, 63
.cgi extension, 36
Chain lengths of carbon atoms, 46–47, 56
Chain saws, 64
Checkerboard design of baseball field, 106
Checksum byte, 77
Chocolate, dogs and, 86
Chromated copper arsenate (CCA), 74
Clapperboard, 7–8
Clock
  atomic, 79–80
  Indiglo watch, 32
Closed captioning, 8
Clothes dryer, electricity consumption of, 62
Coal, electricity generated by, 68–69
Cognitive skills, 99–101
Coils in inductive loops, 51
Coin-operated pool table, 113–14
  two types of cue balls, 114
Cold sensitive nerve endings, 98
Colloids, 90
Color(s)
  of mood ring, 15–16
  of sky, 22
Colorization of black-and-white movies, 18–19
Commercial jets, no locks on doors, 63–64

Compression ratio of the engine, 56
Compression stroke, 48, 56
Computers, 33–43, 62
  brain, 42
  electricity consumption of, 62
  fastest in the world, 41–42
  Linux, 37–38
  RAM, adding, 34–35
  T1 line, 39
  turning off versus leaving on,
    38–39
  Web image formats, 40
  Web page extensions, 35–36
  Year 2038 Problem, 42–43
Concept of reference frames,
  129–30
Cones in eyes, 96, 97
Conversation, decibel rating of, 28
Cool problems, cool solutions,
  119–31
  daytime running lights on cars,
    127–28
  drilling tunnel to the center of
    the earth, 120
  earth's weight, 130
  firing gun on train traveling as
    fast as a bullet, 129–30
  ice-powered air conditioner,
    123–24
  money, 121
  paper production, 124
  satellite viewing, 126
  telescopes detecting equipment
    left behind by astronauts,
    122
  universe matter, 125
Coors Field, 106
Copy paper, carbonless, 17
Cotton candy, 84–85
C programs, 42
Cramping, risk of, 115
Crankshaft, 52–53, 117
  of NASCAR engines, 117
  sump oil system and, 52–53
Crude oil, 46–47
Cubic light year, 125
Cue ball, 113–14

D

Dark blue color of mood ring, 16
Daylight savings time, 79

Daytime running lights on cars,
  127–28
dB. *See* Decibel
Decaffeinated products, labeling
  of, 89
Decibel (db), 28–29
Decompression sickness, 109
Diesel engines, 53–55
  advantages of, 54
  fuel injection of, 53
  historical problems for, 53–54
  waterproofing, 55
Diesel fuel, 46–47
Digestive process, 103, 104, 115
  avoiding swimming after eating,
    115
  flatulence during, 103
  stomach, about, 104
Digital cell phones, 76
Dimples on golf balls, 112
Distance
  effect on intensity of sound, 29
  measuring, 26–27
DLLs. *See* Dynamic link libraries
Dodge Intrepid cars, 116
Dogs, chocolate and, 86
Dolby Laboratories, 7
Doppler shift, 65–66
DOS, 36
Drag of golf balls, 112
Drilling tunnel to the center of
  the earth, 120
Drowning, risk of
  during scuba diving, 109
  swimming after eating, 115
Dry ice, 13–15
  making, 14–15
  safety, 14
Dry sump oil system, 52–53
Dynamic link libraries (DLLs),
  34–35
Dynamite, 70
Dynamometer, 117

E

Ears, sensors in, 97
Earth
  diameter of, 120
  drilling tunnel to the center of, 120
  flashlight beam to moon, 22–23
  weight of, 130–31

Eating, swimming after, 115
Electric blanket, 63
Electricity
  converstion of, 32
  generating from heat, 60
  home appliances' consumption
    of, 62–63
  using solar cells, 66–68
Electric range burner, electricity
  consumption of, 62
Electroluminescence, 32
Electronic ignition of magneto, 64
Email program on computer,
  34–35, 76
Emulsion, 90
Engines
  compression ratio of, 56
  diesel, 53–55
  gasoline versus diesel, 53–55
  magneto to generate power
    for, 64–65
  NASCAR versus regular street
    car, 116–17
  power output, measuring, 117
  turbocharger versus
    supercharger, 47–48
  turning chemical energy to
    mechanical power, 128
  viscosity of oil in, 49
Entertainment, 1–9
  clapperboard, 7
  closed captioning, 8–9
  colorization of black-and-white
    movies, 18–19
  FM radio stations, 9
  fog machines, 3
  light guns, 4–5
  light sabers, 2–3
  rotating camera, 5–6
  sound on motion picture film,
    6–7
Epithelial cells, 104
Equation expressing gravitational
  attraction of two spherical
  objects, 130
ESPN, 110
Ethyl acetate processing of
  caffeine, 88–89
EXE. See Executable file
Executable file (EXE), 34
Exhaust system of turbocharger, 47

Experiments
  gasoline pump, understanding,
    51–52
  hydrogen peroxide on cut
    potato, 95
  making dry ice, 14–15
  popcorn, understanding, 82
  reflection in glass, 50
  sugar in soft drinks, under-
    standing, 83
Extensions on Web pages, 35–36
Eyes
  sensors in, 97
  20/20 vision, 95–96

F

Factory-sealed device, 75
Fatty acids, 103
FCC. See Federal
  Communications Commission
Federal Communications
  Commission (FCC), 9
Federal Reserve, 121
Feedback
  in PA system, 30–31
  on PC, creating, 31
Fescue grass, 106
Fiber optic line, 39
Firecracker, decibel rating of, 28
First-down line, 110–11
Fizzing of Alka-Seltzer, 17
Flashlight beam, 22–23
Flatulence, 103
Flavor-aid, 83
Flour, 61, 86
  explosiveness of, 61
Fluids, viscosity of, 49
FM radio stations, 9
Fog machines, 3
Follicles, hair, 94
Foodstuffs, 12, 81–91
  beer can, widget inside, 87–88
  caffeine, removing from coffee
    beans, 88–89
  carrageenan, 85–86
  chocolate and dogs, 86
  coffee beans, removing caffeine
    from, 88–89
  cotton candy, 84–85
  mayonnaise, 90–91

popcorn, 82
root beer, 83–84
soft drinks, 83
Wint-O-Green Lifesavers, 12
Football, first-down line for, 110–11
Forced induction systems, 47
Frequency Shift Keying (FSK), 77
FSK. *See* Frequency Shift Keying
Fuel oils
  diesel, 46–47, 53
  gasoline, 46–47, 51–57, 127–28
  kerosene, 46–47
  lubricating, 46
  Otto, 69
Fuel injection of diesel engine, 53
Fungicides, 113

G

Gas, body. *See* Flatulence
Gasoline, 46–47, 51–57, 127–28
  consumption, daytime running
    lights, 127–28
  engine versus diesel, 52–55
  octane rating of, 56–57
  pump at filling station, 51–52
Gas station, 51–52, 56–57
G-cells, 104
Gelatin, 86
GIF files, 40
Ginger beer, 83
Glare-resistant setting for rearview
  mirror, 50
Glass and infrared energy, 30
Global positioning satellite
  (GPS), 80
Global warming, 68
Glucose molecules, 103
Golf balls, dimples on, 112
Golf course, grass on the
  greens, 112–13
GPS. *See* Global positioning
  satellite
Grass. *See also* Lawn mower
  on baseball field, 106
  bentgrass, 113
  Bermuda grass, 106, 113
  bluegrass, 106
  fescue, 106
  on golf course, 112–13
  hydroponic system for
    growing, 113

rye grass, 106
zoysia grass, 106
Gravitational attraction of
  two spherical objects,
  equation for, 130
Gray color of mood ring, 16
Green color of mood ring, 16
Greenhouse gas, 68
Gripping the baseball, 107
Growth phase, 94
Guinness beer, 87
Gums, chemical in processed
  foods, 85
Gun
  decibel rating of gunshot, 28
  firing on train traveling as fast as
    a bullet, 129–30
  laser speed, 65–66
  silencer, 27–28

H

Hair dryer, electricity consumption of, 62
Hair on body, 94
Harris-Benedict formula, 102
Hawks' vision, 96
Headlight bulb, watts usage,
  127
Health, 93–104
  calorie intake, 102–3
  flatulence, 103
  hair on body, 94
  hydrogen peroxide, 94–95
  intelligence quotient, 99–101
  knuckle cracking, 98–99
  poison ivy, 96–97
  senses, 97–98
  stomach, 104
  20/20 vision, 95–96
Hearing loss, decibel sound
  causing, 29
Heat
  generating electricity from, 60
  pump, electricity consumption
    of, 62–63
Heat sensitive nerve endings, 98
Helicobacter Pylori, 104
Heptane, 57
Herbicides, 113
Holes in flat prongs on plugs for
  electrical appliances, 75

Hollandaise, 91
Home appliances
  electricity consumption of, 62–63
  flat prongs on plugs, 75
  solar cells for generating, 67
Honey, viscosity of, 49
House, 71–80
  caller ID, 76, 77
  ceiling "popcorn," 73
  cell phones, 76
  electric stud finder, 73–74
  lamps with touch-sensitive
    switches, 78
  plugs, flat prongs on, 75
  pop-up turkey timers, 72
  pressure-treated lumber, 74–75
  radio signals from National
    Atomic Clock, 79–80
  solar panels for, 66–68
.htm extension, 35–36
.html extension, 35–36
Hubble Space Telescope, 23, 122
Humans
  body temperature, touch-sensitive
    switches and, 78
  toxic levels of caffeine for, 86
Humvees, 55
Hydrocarbon molecules of
  petroleum, 46
Hydrochloric acid, 104
Hydrogen, 103
Hydrogen peroxide, 94–95
Hydrogen sulfide, 103
Hydroponic system for growing
  grass, 113

I

Ice, dry, 13–15
Ice-powered air conditioner, 123–24
Ice-skating rinks, 109–10
Igniter on oxygen canister, 62
Ignition key on planes, 63
Image formats on the Web, 40
Immiscible material, 90
Incandescent bulb, 32
Indiglo watch, 32
Inductive loop, 51
Infrared energy, 29–30

Infrared laser light, 66
Intelligence quotient (IQ), 99–101
  increasing, 101
  mental age, 99
  tests designed to measure, 100–101
  value of, 101
Internet service provider (ISP), 39
Iron, 13
ISP. See Internet service provider
Itch sensitive nerve endings, 98

J

Jacobs Vehicle Systems, 48
Jake Brake, 48
Jet engine, decibel rating of, 28
Joints
  popping, 98–99
  sensors in, 98
JPG files, 40

K

Kerosene, 46–47
Kilowatt hours
  calculating, 39
  cost of electricity, 62
Knocking in engine, 56
Knuckle cracking, 98–99
Kola nuts, 88, 89
Kool-aid, 83

L

Lactose, 103
Laminar flow, 112
Lamps with touch-sensitive
  switches, 78
Language ability, IQ tests for, 100
Laser
  beam reflector, 122
  speed gun, 65–66
  versus flashlight, 23
Lawn mower, 28, 64–65, 106
  on baseball fields, 106
  decibel rating of, 28
LCD. See Liquid crystal display
Lead in gasoline, 57
Leak down rate, 117
Leap years, 79
Lecithin, 90
Legal blindness, 96

Lifesavers, Wint-O-Green, 12
Light, triboluminescence, 12
Light and sound, 12, 21–32
  blue light, 12
  decibel, 28–29
  feedback in PA system, 30–31
  flashlight beam from earth to
      moon, 22–23
  gun silencer, 27–28
  Indiglo watch, 32
  light year, 26–27
  motion sensors, 29–30
  sky, blueness of, 22
  sonic boom, 25–26
  zebra's stripes, 24–25
Light bulb
  amount of coal needed to
      power, 68
  electricity consumption of, 62
Light guns on video game, 4–5
Lightning, 12
Light sabers, 2–3
Light year, 26–27, 125
Line designations, 39–40
Lines on ice-skating rinks, 110
Liquid crystal, of mood ring, 15–16
Liquid crystal display (LCD),
    15–16, 78
Liquid nitrogen, 13, 87
  in beer, 87
Locked-out device, 75
Locks on airplanes, 63–64
Locust bean gum, 85
Logos on ice-skating rinks, 110
Lotus, 36
Lubricating oils, 46
Lumber, pressure-treated, 74–75
Lunar Excursion Module, 122

M

Mac OS, 37, 42
Macintosh, 37, 42
Magnesium, 13
Magnetic cue balls, 114
Magneto, 64–65
Magnus, Gustav, 107
Magnus Effect, 107
Mass of the universe, 125
Mathematical ability, IQ
    tests for, 100
Matrix, The, 5

Mayonnaise, 90–91
Memory ability, IQ tests for, 100
Mental retardation, IQ test score
    level for, 100
Methane, 46, 56, 103
Methyl salicylate, 12
Methylene chloride processing of
    caffeine, 88
Microencapsulation technology, 17
Microphone, feedback in PA
    system and, 30–31
Microsoft, 36
Military fatigues, 24
Military vehicles, submersion
    of, 54–55
Milky Way, viewing, 126
Mirror, rearview, 50
Modem, 39, 77, 79
  for caller ID, 77
  over phone line, 79
  T1 line versus residential, 39
Moisture, importance to popcorn
    kernel, 82
Molecules
  crystalline sugar, 12
  glucose, 103
  hydrocarbon, 46
  liquid crystals, 15
Money in the United States, 121
M1 money supply, 121
M2 money supply, 121
M3 money supply, 121
Mood rings, 15–16
  list of colors, 16
Moon
  drilling tunnel to the center of, 120
  flashlight beam from earth, 22–23
  mass of, 120
Morphing, 5
Motion picture films, recording
    sound on, 6–7
Motion sensors, 29–30
Motor oil can, weight of, 49
Movie production
  clapperboard, 7–8
  colorization of, 18–19
  rotating camera around frozen
      actor, 5–6
  Matrix, The, 5
  Star Wars, 2–3
MPG files, 40

Mucosa, 104
Muscles, sensors in, 98

N

Nails, stud finder for, 73–74
Nanosecond, 27, 66
  definition of, 27
NASA, 122
NASCAR engines, 116–117
  alternators, 117
  carburetors, 116
  coolant pumps, 117
  displacement, 116
  ignition system, 116–17
  intake and exhaust, 116
  oil pumps, 117
  radical cam profiles, 116
  running on dynamometer, 117
  steering pumps, 117
National Highway Traffic Safety
  Administration (NHTSA), 127
National Institute of Standards
  and Technology (NIST), 79
Near total silence, decibel rating
  of, 28
Neon light, 32
Nerve endings in skin, 98
Nervous system, 115
Newton, Isaac, 130
NHTSA. *See* National Highway
  Traffic Safety Administration
NIST. *See* National Institute of
  Standards and Technology
Nitrogen, in Guinness beer, 87
Nitrogen narcosis, as danger of
  scuba diving, 109
Nitrogen oxides, 68
Nitroglycerine, 69
Nose, sensors in, 98
Nozzle of gasoline pump, 51–52
.nsf extension, 35–36

O

Octane rating of gasoline, 56–57
Oil
  forms of, 46–47
  pumps, dry sump, 52–53
  viscosity of, 49
On the Go, 45–57
  car's rearview mirror, 50
  diesel engines, 53–55

dry sump oil system, 52–53
  gasoline pump at filling
    station, 51–52
  Jake Brake, 48
  motor oil can, weight of, 49
  octane rating of gasoline, 56–57
  oils, forms of, 46–47
  traffic light sensor, 50–51
  turbocharger versus
    supercharger, 47–48
OPEC oil embargo, 53
Operating system of computer,
  34–35, 37–38
  Linux and, 37–38
  RAM and, 34–35
Otto fuel, 69
Owls' vision, 96
Oxygen canister, 61–62
Oxygen toxicity, as danger of
  scuba diving, 109

P

PA system, feedback in, 30–31
Pain sensitive nerve endings, 98
Paper
  carbonless copy, 17
  production of, 124
Paraffin smoke, 13
Paraffin wax, 47
Parietal cells, 104
Passive InfraRed detector, 29–30
PC. *See* Personal computer
PCS. *See* Personal communication
  services
Pentane, 57
Percent daily values, 102
PERL, 36
Personal communication services
  (PCS), 76
Personal computer, 31, 36
  creating feedback on, 31
Pesticides, 113
Petroleum, 46
Phase diagram, 14
Phone. *See* Telephone
Photodetector, 6
Photodiode, 4
Photons of flashlight beam, 22–23
Photosensor, 29–30
Physical activity, as factor in calcu-
  lating caloric need per day, 102

Pine trees, paper production from, 124
PIR. *See* Passive InfraRed
Pixels, 40–41, 111, 122
.pl extension, 36
Plugs, flat prongs on, 75
Plutonium, 60
Poison ivy, 96–97
Police radar versus laser speed gun, 66
Pollutants, 68
Polymer, 49
Polystyrene, 73
Pool table, coin-operated, 113–14
Popcorn, 82
"Popcorn" on ceiling, 73
Popping of joints, 98–99
Pop Rocks candy, 18
Pop-up turkey timer, 72
Potassium chlorate, 61
Pounds per square inch (PSI)
  diving pressure, 108
  normal pressure for body, 108
  pressure behind bullet, 27–28
  turbocharger of supercharger, 47
Power, 59–70
  coal, 68
  commercial jets, 63–64
  flour, explosiveness of, 61
  generating electricity from heat, 60
  home appliances, 62–63
  laser speed gun, 65–66
  lawn mower, 64–65
  oxygen canister, 61–62
  solar cell, 66–68
  torpedoes, 69–70
Power plant, 60, 68–69
  coal versus nuclear, 68–69
Pressure
  body, normal pressure, 108
  bullet, 27–28
  diving pressure, 108
  -treated lumber, 74–75
Pressure sensitive nerve endings, 98
Propane, 56
Propulsion system of torpedo, 69–70
Propylene glycol dinitrate, 69
PSI. *See* Pounds per square inch

Pumps
  coolant, 117
  gasoline, 51–52
  heat, 62, 63
  oil, 117
  steering, 117
  water pump of well, 62
Pyroelectric sensors, 29–30

R

Radar, 27, 65–66, 74
  laser speed gun, 65–66
  light year and, 27
  in stud finder, 74
Radical cam profiles of NASCAR engines, 116
Radioisotope thermoelectric generators (RTGs), 60
Radio station, frequency of broadcasts, 79
Radio waves, 27
RAM. *See* Random access memory
Random access memory (RAM), 34–35, 37
Rayleigh scattering, 22
Real time, 9
Rearview mirror of car, 50
Refrigerator, electricity consumption of, 62, 63
Rest phase, 94
Revolutions per minute (RPMs), 48, 53
  of diesel engine, 53
  of turbocharger, 48
Rigs, compression release engine braking system of, 48
Rings, mood, 15–16
Rinks, ice-skating, 109–10
Rock concert, decibel rating of, 28
Rods in eyes, 97
Root beer, 83–84
Rotating camera around frozen actor in movies, 5–6
RPMs. *See* Revolutions per minute
RTGs. *See* Radioisotope thermo-electric generators
RVs, compression release engine braking system of, 48
Rye grass, 106

S

Sarsaparilla vine, 83
Sassafras tree, 83
Satellites
  asynchronous orbits of, 126
  global positioning satellite
    (GPS), 80
  power of, 60
  viewing, 126
Scratch-and-sniff stickers, 16–17
"Scratch" in pool, 113
Screen driver electronics, 4
Scripting language, 36
Scuba diving, 108–9
  avoiding the bends, 109
  dangers of, 109
  pressures experienced by diver's
    body, 108
Scuba tanks, 61–62
Seaweed, carrageenan as extract
  from, 86
Self-contained underwater breathing apparatus. See Scuba diving
Senses, 97–98
Sensors
  active, 29
  in coin-operated pool table, 114
  in ears, 97
  in eyes, 97
  in joints, 98
  motion, 29–30
  in muscles, 98
  pyroelectric, 29–30
  traffic light, 50–51
Separation of a laminar boundary
  layer, 112
Serosa, 104
Server side includes (SSI), 36
Shockwave files, 40
.shtml extension, 35–36
Silencer of gun, 27–28
Silly string, 82
Skin
  nerve endings in, 98
  temperature of person, 30
Sky, blueness of, 22
Smog, 68
Snorkel, air intake problem and, 55
Sodium bicarbonate, 17
Sodium chlorate, 61–62

Soft drinks. See also Carbonated
  drinks
  caffeine in, 89
  sugar in, 83
Solar cell, 66–68
Sonic boom, 25–26, 130
Sound. See also Light and sound
  measuring intensity of, 28–29
  motion picture films, 6–7
  travel of, 65
Sound barrier, 25–26
Sound waves, 25, 29, 129–30
Space heater, electricity consump-
  tion of, 62, 63
Space station
  viewing, 126
  oxygen canisters on, 62
Spark plug, generating power
  for, 64–65
Sparks from Wint-O-Green
  Lifesavers, 12
Spatial ability, IQ tests for, 100
Speakers, feedback in PA system
  and, 30
Special effects, 11–19
  Alka-Seltzer fizz, 17
  colorization of black-and-white
    movies, 18–19
  dry ice, 13–15
  mood rings, 15–16
  Pop Rocks candy, 18
  scratch-and-sniff stickers, 16–17
  trick birthday candles, 12
  Wint-O-Green Lifesavers, 12
Speed of light, 27
Spherical objects, gravitational
  attraction of, equation for, 130
Sports, 105–17
  baseball field, checkerboard
    designs on, 106
  coin-operated pool table, 113–14
  curveball in baseball, 107–108
  golf balls, 112
  golf course, 112–13
  ice-skating rinks, 109–10
  NASCAR engines, 116–17
  scuba diving, 108–9
  superimposed first-down
    line, 110–11
  swimming after eating, 115
SporTVision, 110

SSI. *See* Server side includes, 36
Stanford-Binet Intelligence Scale, 99
Standard time library, 42–43
Stars, measuring distance of, 26–27
Star Wars, 2–3
Stern, William, 99
Stickers, scratch-and-sniff, 16–17
Stomach, 104
Stud finder, 73–74
Stuff around the house, 71–80
  caller ID, 76, 77
  ceiling "popcorn," 73
  cell phones, 76
  electric stud finder, 73–74
  lamps with touch-sensitive
    switches, 70
  plugs, flat prongs on, 75
  pop-up turkey timers, 72
  pressure-treated lumber, 74–75
  radio signals from National
    Atomic Clock, 79–80
Sublimation, 13
Submarines
  oxygen canisters on, 62
  torpedoes, 69–70
Sugar
  in cotton candy, 84–85
  in soft drinks, 83
Sulfur dioxide, 68
Sump oil system of car, 52–53
Sun, density of, 125
Sunlight, 22
Supercharger, 47–48, 116
  versus turbocharger, 47–48
Supersonic speeds
  airplanes flying at, 26
  bullets traveling at, 28
Surface temperature of person, 15
Swap file on hard drive, 34
Swimming after eating, 115
Swiss Water Process, 89
Sympathetic nervous system, 115
Synovial fluid, 99

T

Tar, 47
Tartar sauce, 91
Taste, sense of, 97–98
TDMA. *See* Time Division
  Multiple Access

Telephone
  analog versus digital cell phones, 76
  caller ID, 76, 77
  T1 line and phone company, 39–40
Telescope, 23, 26, 122, 126
  detecting light from distant
    stars, 23, 26
  Hubble Space Telescope, 23, 122
  resolving power of, 122
  satellite viewing, 126
Television
  closed captioning for, 8
  superimposing first-down line,
    110–11
Television Decoder Circuitry Act, 8
Temperature
  effect on viscosity, 49
  human body, 78
  skin, of person, 30
  surface, of person, 15
Tetraethyl lead, 57
Thea Sinensis plant, 88
Theobroma Cacao tree, 88
Theobromine, 86
Thermic effect of food, as factor
    in calculating caloric need per
    day, 102–3
Thermocouples, 60
Thermodynamics, laws of, 68
Thermotropic liquid crystals, 15
Thousand Island salad dressing, 91
Time Division Multiple Access
    (TDMA), 76
Timex, 32
T1 line, 39
Tongue, 98
Torpedoes, 69–70
Torvalds, Linus, 38
Touch-sensitive switches, 78
Traffic light sensor, 50–51
Trees, paper production from, 124
Triboluminescence, definition
    of, 12
Trick birthday candles, 12
Trimmers, 64
Turbine, 47, 60
  generating electricity from heat, 60
  of turbocharger, 47
Turbocharger, 47–48, 116
  versus supercharger, 47–48
Turbulent boundary layer, 112

Turkey timer, pop-up, 72
20/20 vision, 95–96
2038 Problem 42–43

U

Ulcers, 104
Ultrasonic sound waves, 29
Ultraviolet light, 12
Universe
  mass of the, 125
  matter in the, 125
UNIX, 36, 37–38
URLs, common extensions on the
  end of, 35
Urushiol, 96–97
U-shaped armature, 64–65

V

V-chip, 8
Vehicles, waterproofing, 55. *See
  also* Automobiles; Engines
Venturi, 52
Vermiculite, 73
Video game, light guns on, 4–5
Virtual Basic code, 36
Virtual memory manager
  (VMM), 34
Viscosity of oil, 49
Visible light versus ultraviolet
  light, 12
Vision
20/20, 95–96
  sensors of, 97
VMM. *See* Virtual memory
  manager

W

Wake, definition of, 25
Watch
  Indiglo, 32
  receiving signals from the
    National Atomic Clock, 79–80
Water
  density of, 125
  solid-liquid-gas behavior of, 14
  viscosity of, 49
Water heater, electricity consumption of, 62, 63
Water processing of caffeine, 88–89

Waterproofing vehicles, 55
Water pump of well, electricity
  consumption of, 62
Wavelength, 22
Waves
  radio, 27
  Sound, 25, 29, 129–30
  water, 25
Web, the
  image formats on, 40
  page extensions, 35–36
Weight on motor oil can, 49
Wet sump oil system, 52–53
Whisper, decibel rating of, 28
Widget inside beer can, 87–88
Windows operating system, 36,
  37–38, 42
  Linux and, 37–38
  speed of, 42
Web pages and, 36
Wintergreen flavoring, 12
Wint-O-Green Lifesavers, 12
Wood, pressure-treated, 74–75
WWVB radio station, 79

XYZ

Xanthan gum, 85
Y2K problem, 42, 43
Year 2038 Problem on computers,
  42–43
Yeast in beer, 83–84
Zebra's stripes, camouflage of,
  24–25
Zoysia grass, 106